cosmetics

Plant Extracts in Skin Care Products

Edited by
Beatriz P. P. Oliveira and Francisca Rodrigues
Printed Edition of the Special Issue Published in *Cosmetics*

www.mdpi.com/journal/cosmetics

MDPI

Plant Extracts in Skin Care Products

Plant Extracts in Skin Care Products

Special Issue Editors

Beatriz P.P. Oliveira
Francisca Rodrigues

MDPI • Basel • Beijing • Wuhan • Barcelona • Belgrade

MDPI

Special Issue Editors

Beatriz P. P. Oliveira
LAQV/REQUIMTE—Faculty of Pharmacy,
University of Porto
Portugal

Francisca Rodrigues
LAQV/REQUIMTE—Faculty of Pharmacy,
University of Porto
Portugal

Editorial Office
MDPI
St. Alban-Anlage 66
Basel, Switzerland

This is a reprint of articles from the Special Issue published online in the open access journal *Cosmetics* (ISSN 2079-9284) from 2017 to 2018 (available at: http://www.mdpi.com/journal/cosmetics/special_issues/plant_extracts)

For citation purposes, cite each article independently as indicated on the article page online and as indicated below:

LastName, A.A.; LastName, B.B.; LastName, C.C. Article Title. *Journal Name* **Year**, *Article Number*, Page Range.

ISBN 978-3-03897-160-3 (Pbk)
ISBN 978-3-03897-161-0 (PDF)

Contents

About the Special Issue Editors

Beatriz P.P. Oliveira is Associate Professor with Habilitation from University of Porto, Faculty of Pharmacy, Director of the Master Course in Quality Control, member of the EFSA Scientific Network for Risk Assessment of GMOs, member of a specialized committee from ASAE. She is the Head of GRESA, a REQUIMTE/LAQV research group which includes PhD researchers, PhD, and MSc students. She has a large experience in student supervision (19 PhD and 80 MSc). She co-authored more than 350 scientific publications in international journals, besides 70 publications in national journals, 40 book chapters, and 900 communications in national and international scientific congresses. Her achievements include, among others , the evaluation of bioactive compounds in plant extracts and agro-industrial by-products, resulting in the submission of four patents. Food security and environmental sustainability are other lines of research that she is pursuing. She has developed several projects with companies.

Francisca Rodrigues, PhD, is a graduate in Pharmaceutical Sciences at Faculty of Pharmacy of University of Porto (UP) in 2008, and finished a Master in Quality Control at UP in 2011. Afterwards, she followed her post-graduate studies and in May 2016 finished her PhD in Pharmaceutical Technology in UP. In the meanwhile, she work with the regulatory department of cosmetic and medical devices companies where acquired a deep experience on the field of regulation and cell culture and in vivo human studies. Her current research is focused on the study of new cosmetic ingredients based on food by-products and their application in cosmetics. In particular, the in vitro, ex vivo and in vivo evaluation and characterization of different by-products with interest in skin aging. So far, she has co-supervised 8 MSc students (4 completed) and 1 PhD student. She published 30 papers in international peer reviewed (ISI) journals and 10 book chapters.

Preface to "Plant Extracts in Skin Care Products"

This book provides readers with a fundamental understanding of the science and applications of plant materials as active ingredients in cosmetics. The chapters of this handbook cover the basics of ethnobotany, (bio)active compounds, and their natural sources. An assessment of the safety (in vitro and in vivo) of some of these plants and preparations is also included. Readers will also learn about concepts central to quality control processes, sustainable management, wild harvesting, and the economic valuation of plants. The volume contains 3 review papers and 7 research manuscripts based on the evaluation of plants for the cosmetic industry.

This book is intended as a handbook for undergraduate and graduate students, teaching professionals in research and higher education institutions, involved in pharmacy, natural products, and chemistry, and cosmetic industry professionals. Non-experts interested in natural and sustainable cosmetics will also find the content informative.

Beatriz P.P. Oliveira, Francisca Rodrigues
Special Issue Editors

Communication

Toxic Evaluation of *Cymbopogon citratus* Chemical Fractions in *E. coli*

Fabiana Fuentes-León [1,*], Maribel González-Pumariega [1], Marioly Vernhes Tamayo [2], Carlos Frederico Martin Menck [3] and Ángel Sánchez-Lamar [1,*]

[1]   Plant Biology Department, Faculty of Biology, University of Havana, Havana 10400, Cuba; bel@nauta.cu
[2]   Radiobiology Department, Center of Applied Technology and Nuclear Development (CEADEN), Havana 10400, Cuba; mariolys@ceaden.edu.cu
[3]   Microbiology Department, Instituto de Ciências Biomédicas, Universidade de São Paulo, São Paulo 05508-900, Brazil; cfmmenck@usp.br
*   Correspondence: fabiana@fbio.uh.cu (F.F.-L.); alamar@fbio.uh.cu (A.S.-L.); Tel.: +53-7832-8542 (F.F.-L. & A.S.-L.)

Received: 2 May 2017; Accepted: 14 June 2017; Published: 17 June 2017

Abstract: *Cymbopogon citratus* (DC) Stapf is consumed as a popular decoction owing to its nice flavor and hypotensor property. Its aqueous extract radioprotector and antimutagenic properties have been experimentally demonstrated. In addition, its DNA protective activity against UV light has been proved in plasmid DNA and bacterial models. The fractioning process is important in order to identify phytocompounds responsible for this activity. In this work, the toxicity of three fractions obtained from *Cymbopogon citratus* (essential oils, butanolic and aqueous fractions) were tested using the SOS Chromotest in *Escherichia coli*. *Cymbopogon citratus* chemical fractions possess cytotoxic properties in *E. coli* in the following order butanolic > aqueous > essentials oils. Genotoxic properties were detected in any of the fractions.

Keywords: essential oils; butanolic and aqueous fractions; SOS assay

1. Introduction

The therapeutic use of plants is part of universal human culture. Over the last few years, there has been an increase in the number of studies reporting medicinal plants and dietary components as chemopreventive agents. Particularly, the usefulness of plants as sources of photochemopreventive components has been demonstrated [1]. For the last 10 years, our research group has been evaluating plant extracts against DNA damage induced by UV light [2–8]. The study of photoprotective properties of phytochemicals enriches and supports the continuous development of the pharmaceutical and cosmetic industries. To achieve this purpose, the assessment of plant extracts' safety, as well as plant-derived products, is highly relevant [9,10].

Cymbopogon citratus (DC) Stapf, known in Cuba as Caña Santa, is consumed as a popular decoction. This plant has been traditionally used to treat different diseases and has several pharmacological properties [11,12]. In addition, antimutagenic properties have been described for this plant, including photoprotective activity evaluated in plasmid DNA, bacterial cells, and eukaryotic organisms [4–6,13,14]. Plenty of these properties have been assayed in total extracts or decoctions, although less of them have been tested in fractioned extracts. The fractioning process is important to identify the compounds responsible for photoprotective activity. Also, the assessment of their safety is a prerequisite in antigenotoxic evaluations.

SOS Chromotest has been used for several years to test different mutagenic agents [15,16]. More recently, it has been very useful in evaluating plant extracts' genotoxicity and DNA protection [17].

In this paper, using a fluorescent protocol of SOS Chromotest in *E. coli* [18], we studied the toxic activity of essential oils, as well as butanolic and aqueous fractions of *C. citratus*.

2. Materials and Methods

2.1. Chemical Fractioning of Cymbopogon citratus Total Extract

Leaves of Caña Santa were obtained from adult plants grown in Boyeros, Havana, Cuba. The specimens were verified at the Medicinal Plants Station in Güira de Melena, Artemisa, Cuba (herbarium # 4593) [19]. Fresh leaves were triturated and boiled in distilled water (w/v) for 30 min. Essential oils were collected by the steam distillation method [20]. The remaining solution was fractionated by extracting in butanol (Sigma, St. Louis, MO, USA) [21]. The resulting aqueous phases were dried and vaporized to obtain butanolic and aqueous fractions. The stock solutions of oil and butanolic fraction (16 mg/mL) were diluted in Dimethyl Sulfoxide (DMSO) (Merck, Darmstadt, Germany) at 2.0 and 2.8%, respectively, and miliQ water. The aqueous fraction was diluted only in miliQ water. The concentrations evaluated were 0.1, 0.5, 1.0, 2.0, and 4.0 mg/mL.

2.2. Bacterial Strains and Culture

E. coli PQ37 strain genotype (F thr leu his-4 pyrD thi gal; K o galT lac ΔU 169 sr/300::Tn10 rpoB rpsL uvrA trp::muc+ sfiA::mud(ap,lac) cts) was used in the SOS Chromotest. The cells were grown at 37 °C under constantshaking (100 rev/min) in Luria-Bertani (LB) (Sigma) media supplemented with Ampicillin (Sigma) (25 μg/mL) until an optic densitometry (OD) of 0.4 at 600 nm.

2.3. SOS Chromotest

The fluorescent SOS assays described by Cuétara, et al. [18] was used to test the toxicity of the chemical fractions. Briefly, exponential phase cultures (OD_{600nm} = 0.4) were 10-fold diluted in fresh LB medium (2X) supplemented with Ampicillin 25 μg/mL and dispensed in 1.5-mL tubes containing the fraction to be tested. The cells were exposed for 30 min at 4 °C. Later, cells were incubated for 2 h at 37 °C under constant shaking (100 rpm/min), and finally the enzymatic reactions were conducted. The cytotoxicity criterion was the reduction of the alkaline phosphatase expression. For the genotoxicity test, the criterion was the increase of the induction factor of the SOS response (SOSIF) [15], calculated as follows:

$$\text{SOSIF} = \frac{(\beta\text{galactosidase}/\text{alkaline phosphatase})_{\text{treated cells}}}{(\beta\text{galactosidase}/\text{alkaline phosphatase})_{\text{non−treated cells}}} \tag{1}$$

Cells harvested in medium without irradiation and irradiated were used as negative and positive controls, referred as 0.0 concentration and UVC (ultraviolet light, band C), respectively. After 30 min of incubation at 4 °C, a 1.5-mL batch was irradiated in Petri dishes with a diameter of 3 cm. UVC irradiation (λ = 254 nm, E = 45 J/m^2) was carried out using a Vilber Lourmat Lamp T15M 15 W (Vilber Lourmat, Suebia, Germany) at room temperature. Afterwards, cells were collected by centrifugation, resuspended in their respective treatment, and then incubated for 2 h at 30 °C under constant shaking (100 rpm/min). All measurements were taken in triplicate.

2.4. Statistical Analysis

The means and standard deviation of alkaline phosphatase and SOSIF were determined. Controls and treatments were analyzed using the Kolmogorov-Smirnov test. Variance homogeneity (Brown Forsythe test) and single classification ANOVA were also conducted. Values for different treatments were compared with the negative control using a Dunett test ($p < 0.05$), according to STATISTICA 6.0.

3. Results

The toxicity of chemical fractions of *C. citratus* to *E. coli* PQ37 cells was investigated. The alkaline phosphatase assay in treated cells (protein synthesis inhibition indices) was used as the cytotoxicity criterion. The capability to produce primary DNA damage by the plant fractions was determined measuring the induction of SOS genes. It is known that a compound is classified as not genotoxic if the SOSIF remains ≤1.5, not conclusive if SOSIF is between 1.5 and 2.0, and genotoxic if SOSIF exceeds 2.0 [22].

The results showed that the essential oils (4.0 mg/mL), aqueous fraction (1.0–2.0 mg/mL), and butanolic fraction (0.5–4.0 mg/mL) decrease the alkaline phosphatase constitutive expression when they are compared to non-treated cells (0.0 mg/mL). Any fraction induced DNA damage at the concentrations tested according to the criterion described by Kevekores et al. [22] (Table 1).

Table 1. Effects of essential oils, and aqueous and butanolic fractions from *Cymbopogon citratus* on the alkaline phopatase (AP) expression and induction of SOS phenomenon (SOSIF) in *E. coli* cells.

	Chemical Fractions					
Concentrations (mg/mL)	Essential Oils		Aqueous		Butanolic	
	AP	SOSIF	AP	SOSIF	AP	SOSIF
0.0	6.34 ± 0.13	1.00 ± 0.05	3.73 ± 0.16	1.02 ± 0.02	3.98 ± 0.32	1.02 ± 0.05
0.1	4.87 ± 0.28	0.98 ± 0.09	3.69 ± 0.13	1.02 ± 0.03	3.66 ± 0.26	1.00 ± 0.06
0.5	6.67 ± 0.96	0.88 ± 0.12	3.40 ± 0.14	1.01 ± 0.03	3.15 ± 0.11 *	1.00 ± 0.03
1.0	6.79 ± 0.68	0.94 ± 0.09	3.43 ± 0.19	1.11± 0.04	2.91 ± 0.29 *	0.93 ± 0.02
2.0	5.00 ± 0.15	0.80 ± 0.05	3.21 ± 0.14 *	1.12 ± 0.04	2.74 ± 0.05 *	1.01 ± 0.05
4.0	3.68 ± 0.12 *	0.90 ± 0.12	2.94 ± 0.07 *	1.34 ± 0.12 *	2.75 ± 0.03 *	1.12 ± 0.07
UVC (5 J/m²)	4.32 ± 0.09 *	7.85 ± 0.84 *	3.11 ± 0.13 *	7.69 ± 0.74 *	2.15 ± 0.23 *	6.55 ± 0.30 *

(*) significant in Dunett Test $p < 0.05$.

4. Discussion

In our work, the fractions obtained from *C. citratus* were cytotoxic in *E. coli* cells PQ37 strain. The butanolic fraction was the most cytotoxic. The essentials oils and aqueous fraction were toxic at the highest concentrations tested. On the other hand, any fractions caused damage in bacterial DNA. In previous experiments in *Caulobacter crescentus* model, the essential oils of *C. citratus* also showed cytotoxic effect at 4.0 mg/mL [6]. Interestingly, in contrast with our results here, the aqueous fraction did not show a toxic result at any concentration tested, and the butanolic fraction showed a strongly genotoxic result [6]. Probably the genetic background of *C. crescentus* and *E. coli* used in our studies could be related to the differences observed. Furthermore, transcriptional fusion in *E. coli* PQ37 and *C. crescentus* NA1000 pp3213 could reinforce this hypothesis [15,16,23].

Using different extraction methods, it was demonstrated that the aqueous extract of *Cymbopogon citratus* did not induce DNA primary in the plasmidic model, and in *E. coli* PQ-37 cells reduced alkaline phosphates levels at 4.0 mg/mL. In this bacterial cell, concentrations higher than 2.0 mg/mL were genotoxic [4,5,18,23].

It is important to remark that in the fractioning process, the phytocompounds of the total extract are separated and concentrated. These aspects could explain the increase in toxicity found in chemical fractions in contrast to the total extract when the same concentrations are tested.

The results obtained here complement previous evaluations of the total extract and chemical fractions [4–6]. Although complementary tests, such as in vitro human epidermis models, should be conducted in order to guarantee the safety of these fractions, these results are an important contribution to the future studies of *Cymbopogon citratus* fractions as chemopreventive agents. Furthermore, nontoxic concentrations bring a good perspective to their potential in the pharmaceutical and cosmetic industries.

5. Conclusions

The essential oils, aqueous and butanolic fractions obtained from *Cymbopogon citrates* possess cytotoxic effects in *E. coli* cells. However, non genotoxic effects were detected at the assayed concentrations.

Acknowledgments: CAPES (Brazil)-MES (Cuba) collaborative project financed this work.

Author Contributions: Ángel Sánchez-Lamar, Carlos Frederico Martin Menck and Maribel González-Pumariega conceived and designed the experiments; Maribel González-Pumariega and Marioly Vernhes Tamayo performed the experiments; Fabiana Fuentes-León analyzed the data; Ángel Sánchez-Lamar, Carlos Frederico Martin Menck and Marioly Vernhes Tamayo contributed reagents/materials/analysis tools; Fabiana Fuentes-León wrote the paper.

Conflicts of Interest: The authors declare no conflict of interest.

References

1. Rojas, J.; Londoño, C.; Ciro, Y. The health benefits of natural skin UVA photoprotective compounds found in botanical sources. *Int. J. Pharm. Pharm. Sci.* **2016**, *8*, 13–23.
2. Vernhes, M.; González-Pumariega, M.; Schuch, A.P.; Menck, C.F.M.; Sánchez-Lamar, A. El extracto acuoso de *Phyllanthus orbicularis* K protege al ADN plasmídico del daño inducido por las radiaciones ultravioletas. *ARS Pharm.* **2013**, *54*, 16–23.
3. Vernhes, M.; González-Pumariega, M.; Andrade, L.; Schuch, A.P.; Lima-Bess, K.M.D.; Menck, C.F.M.; Sánchez-Lamar, A. Protective effect of a *Phyllanthus orbicularis* aqueous extract against UVB light in human cells. *Pharm. Biol.* **2013**, *51*, 1–7. [CrossRef]
4. Vernhes, M.; González-Pumariega, M.; Fuentes-León, F.; Baly-Gil, L.; Menck, C.F.M.; Sánchez-Lamar, A. Desmutagenic activity of *Cymbopogon citratus* and *Phyllanthus orbicularis* against UVC damage in *E. coli*. *Adv. Pharm. J.* **2016**, *1*, 80–85.
5. González-Pumariega, M.; Fuentes-León, F.; Vernhes, M.; Schuch, A.P.; Menck, C.F.M.; Sánchez-Lamar, Á. El extracto acuoso de *Cymbopogon citratus* protege al ADN plasmídico del daño inducido por radiación UVC. *ARS Pharm.* **2016**, *57*, 193–199.
6. Fuentes-León, F.; García-Fernández, F.; Aguilera-Roque, K.B.; Menck, C.F.M.; Galhardo, R.d.S.; Sánchez-Lamar, Á. Fotoprotección al DNA ejercida por fracciones químicas obtenidas de *Cymbopogon citratus* (DC) Stapf. *Rev. Cubana Cienc. Biol.* **2017**, in press.
7. Vernhes, M.; Schuch, A.P.; Fuentes-León, F.; Menck, C.F.M.; Sánchez-Lamar, Á. Extracto acuoso de *Pinus caribaea* inhibe el daño inducido por radiaciones ultravioletas, en DNA plamídico. *J. Pharm. Pharmacogn. Res.* **2017**, *5*, 262–269.
8. Menéndez-Perdomo, I.M.; Fuentes-León, F.; Wong-Guerra, M.; Carrazana, E.; Casadelvalle, I.; Vidal, A.; Sánchez-Lamar, Á. Antioxidant, photoprotective and antimutagenic properties of *Phyllanthus* spp. from Cuban flora. *J. Pharm. Pharmacogn. Res.* **2017**, *5*, 251–261.
9. Seebode, C.; Lehmann, J.; Emmert, S. Photocarcinogenesis and skin cancer prevention strategies. *Anticancer Res.* **2016**, *36*, 1371–1378.
10. Saewan, N.; Jimtaisong, A. Natural products as photoprotection. *J. Cosmet. Dermatol.* **2015**, *14*, 47–63. [CrossRef]
11. Ekpenyong, C.E.; Akpan, E.; Nyoh, A. Ethnopharmacology, phytochemistry, and biological activities of *Cymbopogon citratus* (DC) Stapf extracts. *Chin. J. Nat. Med.* **2015**, *13*, 0321–0337. [CrossRef]
12. Nambiar, V.S.; Matela, H. Potential functions of lemon grass (*Cymbopogon citratus*) in health and disease. *Int. J. Pharm. Biol. Arch.* **2012**, *3*, 1035–1043.
13. Cápiro, N.; Sánchez-Lamar, A.; Baluja, L.; Sierra, L.M.; Comendador García, M.A. Efecto de la concentración de *Cymbopogon citratus* (DC) Stapf sobre la genotoxicidad de mutágenos modelos, en el Ensayo Smart de Ojos W/W+ de *Drosophila melanogaster*. *Rev. CENIC Cienc. Biol.* **2005**, *36*, 1–5.
14. Cápiro, N.; Sánchez-Lamar, Á.; Fonseca, G.; Baluja, L.; Borges, E. Capacidad protectora de *Cymbopogon citratus* (DC.) Stapf. ante el daño genético inducido por estrés oxidativo. *Rev. Cuba. Investig. Biomed.* **2001**, *20*, 33–38.

15. Quillardet, P.; Frelat, G.; Nguyen, V.D.; Hofnung, M. Detection of ionizing radiations with the SOS Chromotest, a bacterial short-term test for genotoxic agents. *Mutat. Res.* **1989**, *216*, 251–257. [CrossRef]
16. Quillardet, P.; Hofnung, M. Induction by UV light of the SOS function sfiA in *Escherichia coli* strains deficient or proficient excision repair. *J. Bacteriol.* **1984**, *157*, 35–38. [CrossRef]
17. Fuentes, J.L.; Alonso, A.; Cuétara, E.; Vernhes, M.; Álvarez, N.; Sánchez-Lamar, Á.; Llagostera, M. Usefulness of the SOS Chromotest in the study of medicinal plants as radioprotectors. *Int. J. Radiat. Biol.* **2006**, *82*, 323–329. [CrossRef]
18. Cuétara, E.; Álvarez, A.; Alonso, A.; Vernhes, M.; Sánchez-Lamar, A.; Festary, T.; Rico, J. A microanalytical variant of the SOS Chromotest for genotoxicological evaluation of natural and synthetic products. *Biotecnol. Apl.* **2012**, *29*, 108–112.
19. Martínez-Guerra, J.M.; Betancour-Badell, J.; Ramírez-Albajes, A.R.; Barceló-Pérez, H.; Meneses Valencia, R.; Lainez-Azcuy, A. Evaluación toxicológica aguda de los extractos fluidos al 30 y 80% de *Cymbopogon citratus* (D.C.) Stapf (Caña Santa). *Rev. Cuba. Plantas Med.* **2000**, *5*, 97–101.
20. Elhassan, I.A.; Eltayeb, I.M.; Khalafalla, E.B. Physiochemical investigation of essential oils from three *Cymbopogon* species cultivated in Sudan. *J. Pharmacogn. Phytochem.* **2016**, *5*, 24–29.
21. Emrizal, A.F.; Yuliandari, R.; Rullah, K.; Indrayani, N.R.; Susanty, A.; Yerti, R.; Ahmad, F.; Sirat, H.M.; Arbain, D. Cytotoxic activities of fractions and two isolated compounds from sirih merah (Indonesian red betel), *Piper crocatum* Ruiz & Pav. *Procedia Chem.* **2014**, *13*, 79–84.
22. Kevekordes, S.; Mersch-Sundermann, V.; Burghaus, C.; Spielberger, J.; Schmeiser, H.; Arlt, V.; Dunkeberg, H. SOS induction of selected naturally occurring substances in *Escherichia coli* (SOS chromotest). *Mutat. Res.* **1999**, *445*, 81–91. [CrossRef]
23. Galhardo, R.S.; Rocha, R.P.; Marques, M.V.; Menck, C.F.M. An SOS-regulated operon involved in damage-inducible mutagenesis in *Caulobacter crescentus*. *Nucleic Acids Res.* **2005**, *5*, 2603–2614. [CrossRef]

Review

Revisiting Amazonian Plants for Skin Care and Disease

Bruno Burlando [1,2] **and Laura Cornara** [2,3,*]

1 Department of Pharmacy, University of Genova, Viale Benedetto XV, 16132 Genova, Italy; burlando@difar.unige.it
2 Biophysics Institute, National Research Council (CNR), via De Marini 6, 16149 Genova, Italy
3 Department of Earth, Environment and Life Sciences, University of Genova, Corso Europa 26, 16132 Genova, Italy
* Correspondence: cornara@dipteris.unige.it; Tel.: +39-010-2099364

Received: 1 July 2017; Accepted: 20 July 2017; Published: 26 July 2017

Abstract: This review concerns five species of trees and palm trees that occur as dominant plants in different rainforest areas of the Amazon region. Due to their abundance, these species can be exploited as sustainable sources of botanical materials and include *Carapa guianensis* Aubl., family Meliaceae; *Eperua falcata* Aubl., family Fabaceae; *Quassia amara* L., family Simaroubaceae; and *Attalea speciosa* Mart. and *Oenocarpus bataua* Mart., family Arecaceae. For each species, the general features, major constituents, overall medicinal properties, detailed dermatological and skin care applications, and possible harmful effects have been considered. The major products include seed oils from *A. speciosa* and *C. guianensis*, fruit oil from *O. bataua*, and active compounds such as limonoids from *C. guianensis*, flavonoids from *E. falcata*, and quassinoids from *Q. amara*. The dermatologic and cosmetic applications of these plants are growing rapidly but are still widely based on empiric knowledge. Applications include skin rehydration and soothing; anti-inflammatory, antiage, and antiparasite effects; hair care; burn and wound healing; and the amelioration of rosacea and psoriasis conditions. Despite a limited knowledge about their constituents and properties, these species appear as promising sources of bioactive compounds for skin care and health applications. An improvement of knowledge about their properties will provide added value to the exploitation of these forest resources.

Keywords: Amazonian tree species; antiage properties; essential fatty acids; flavonoids; hair care; humectant; limonoids; quassinoids; skin soothing; wound healing

1. Introduction

The search for new bioactive principles to be used in pharmaceutical and cosmetic products is to a wide extent directed towards natural sources, mostly botanical entities. The Amazon region holds extraordinarily rich plant diversity, and therefore it is attracting much interest for the discovery of new bioactive principles. Traditional plant remedies are generally a preferential starting point for projects of drug discovery from natural sources. However, unlike Asian traditional medicines, popular remedies from the Amazon have not been recorded for thousands of years in herbal pharmacopoeia. By contrast, indigenous groups have made extensive use of rainforest plant materials to meet their health needs. Only in recent times, the Amazonian ethnobotanical culture has started to be discovered and taken into consideration for the development of new drugs and skin care products.

In this review, we report five species of trees and palm trees that occur as dominant species in different rainforest areas of the Amazon region. Due to their abundant occurrence and the possibility of cultivation, these species can be exploited as sustainable sources of botanical materials. They include *Carapa guianensis* Aubl., a timber tree of the Meliaceae family that can be cultivated and is known as alternative mahogany; *Eperua falcata* Aubl., a timber tree of Fabaceae that is dominant in Guyana forests; *Quassia amara* L., a small tree belonging to Simaroubaceae that is cultivated on a commercial scale; and two very abundant Arecaceae species, *Attalea speciosa* Mart. and *Oenocarpus bataua* Mart. Each species has been treated by considering its general features, major constituents, overall medicinal properties, detailed dermatological and skin care applications, and any harmful effects. Literature data on the medicinal uses of plants and their constituents have been collected from different online databases, including Scopus, Web of Science, PubMed, Google Scholar, https://clinicaltrials.gov, and the Espacenet patent repository.

2. *Attalea speciosa* Mart.

Family: Arecaceae; synonym: *Orbignya phalerata* Mart.; common name: babassu; parts used: seeds and fruits. INCI (International Nomenclature of Cosmetic Ingredients) names [1]: Attalea Speciosa Mesocarp Extract and Attalea Speciosa Seed Oil.

2.1. Features

This is a palm tree reaching 30 m in height. The stem carries at the apex a crown of large, pinnated leaves; the flowers are assembled in large, axillary, bending inflorescences up to 1.5 m long; and the fruits are oblong nuts of about 6 cm, rusty in color and containing two to six seeds. The species is widespread and economically important, especially in Maranhão, a state in northeastern Brazil. The fruit is exploited as an energy source, food, or medicine, while the seeds obtained by manual breaking of the fruit are transformed into oil [2–4].

2.2. Constituents

The seeds contain about 65% to 68% lipids. The oil is similar to coconut oil and is composed mainly of triglycerides with monounsaturated and saturated fatty acids. The major oil constituents include lauric (about 50%), myristic (20%), palmitic (11%), oleic (10%), stearic (3.5%), and linoleic (1.5%) acids [5]. Due to its high degree of saturation, the oil is semi-solid at 20 °C, while it melts completely at temperatures above 25 °C to 30 °C.

An ethanol extract of the leaves was reported to contain flavonoids, steroids, and/or triterpenoids, and saponins, while gas chromatography analysis showed linolenic acid as a major constituent, followed by the terpene citronellol and the fatty acids linoleic, palmitic, capric, and stearic acid [6].

2.3. Properties

Popularly used remedies in Maranhão, Brazil, are the fruit mesocarp, oil, and a residue of oil production called 'borra'. Mesocarp flour is used for gastritis, inflammation, and leucorrhea; the seed residue is used for wound healing; and the oil is used for wounds and leucorrhea [7]. In addition, the fruit is used for pain and rheumatism, constipation, obesity, leukemia, and circulation, while the oil is also used as an antimicotic and a laxative [8]. Experimental studies in vivo and on tumor cell lines have strengthened the idea that the plant could be useful for venous dysfunctions and as an adjuvant in antitumor treatments [9].

2.4. Dermatologic and Cosmetic Uses

In a study conducted on rats, an aqueous extract of the fruit has induced positive effects on wound healing [10].

The oil has been known as a skin care remedy for centuries because it is emollient and penetrates easily through the skin without leaving a greasy feeling. The oil can be used in cosmetics as an alternative to coconut oil to treat dry skin, itchiness, eczema, and various irritations. The two main oil constituents, namely lauric and myristic acids, have a melting point that is close to body temperature. Consequently the oil melts completely when coming to contact with skin, thus absorbing heat and inducing a cooling effect that corroborates its emollient virtues [11,12]. The oil is also suitable for dry, dull hair as it confers volume by avoiding oily appearance [13]. Populations inhabiting the Tucuruí Lake Protected Areas in the eastern Amazon use the oil for skin infections, myiasis, and mycosis. In addition, they use it as hair moisturizing agent, and, curiously, the oil from the beetle *Pachymerus nucleorum* larva, which develops inside the seed of the coconut, is also used for the same purpose [4].

2.5. Adverse Effects

An extract in water of the pulverized mesocarp of the fruit, administered to mice at doses of up to 3 g/kg did not induce harmful effects [14]. Moreover, no adverse effects are known in humans due to the use of oil or other products of the plant.

3. *Carapa guianensis* Aubl.

Family Meliaceae; common name: andiroba; parts used: seeds, leaves, and bark. INCI names: Carapa Guaianensis Oil PEG-8 Esters, Carapa Guaianensis Oil Polyglyceryl-6 Esters, Carapa Guaianensis Seed Oil, and Carapa Guaianensis Seed Powder.

3.1. Features

This is a large tree reaching 30 m in height, with a cylindrical trunk and a thick, oval crown. The bark is grey or brown, with a fissurated and scaly appearance. The leaves are paripinnate and alternate, with elliptic-to-lanceolate leaflets, dark green above and opaque below. The flowers are clustered in large, unisexual inflorescences, and the fruit is a dehiscent capsule, dropping at maturity and containing various seeds [15,16].

The species grows both wild and cultivated in equatorial and tropical zones of South America. The wood has commercial value and is known as 'Brazilian mahogany'. The seeds are used to extract a pale yellow oil that is economically important in the Amazon region [17].

3.2. Constituents

The seed oil is rich in essential fatty acids, prevalently saturated ones, and melts at a temperature of about 25 °C. The main components include oleic, palmitic, linoleic, myristic, and stearic fatty acids [18], while other components include squalene, stigmasterol, and cholesterol [19]. The seeds and other plant portions contain limonoids (tetranortriterpenoids) such as methyl-angolensate, gedunin, 7-deacetoxy-7-oxogedunin, deacetylgedunin, 6α-acetoxygedunin, and andirobin. Phragmalin-type and mexicanolide-type limonoids, collectively known as carapanolides, have also been isolated from seeds and oil [20]. Limonoids confer to the plant and the oil a bitter taste [21], from which derives the name 'andiroba' in the Tupi-Guarani language [22]. Other compounds include triterpenes (e.g., ursolic acid), flavonoids (naringenin), coumarins (scopoletin), benzoic acid derivatives (3,4- and 2,6-dihydroxymethylbenzoate), and long-chain fatty acids (tetra-triacontanoic, triacontanoic acids) [23]. Flowers and leaves contain volatile compounds such as the cyclic sesquiterpenes bicyclogermacrene, germacrene B, germacrene D, and α-humulene [24].

3.3. Properties

People in the Amazon region have been using the plant as traditional medicine for centuries. Its reported uses include bark tisanes against intestinal ailments, parasites, and skin problems, and, in addition, an oil-derived soap used for its anti-inflammatory, antimicrobic, antiarthritic effects and as an insect repellent [25]. Guyana Patamona people macerated bark in water and used it for eczema, measles, and chicken pox, an infectious disease causing mild fever and inflamed blisters [26].

The plant anti-inflammatory properties are thought to depend on the presence of limonoids, as testified by studies on murine models of arthritis and allergy [27,28]. Edema formation has been inhibited by the oil in different rodent models via the impairment of signaling pathways triggered by histamine, bradykinin, and platelet-activating factor [29]. Other studies have shown toxic properties of limonoids on insect larvae [30,31]. More specifically, gedunin and some of its derivatives have been investigated for antimalarial, hepatoprotective, and antitumor properties [32–34]. Anti-inflammatory, analgesic, and antiallergic virtues of the oil have led to the development of pharmaceutical products [35].

3.4. Dermatologic and Cosmetic Uses

In traditional medicine, the oil is applied to wounds or used for massage, as insect repellent, and for skin problems like psoriasis. It is also popularly used to prevent sarcopsyllosis and pediculosis [36].

An emulsion of the oil and the synthetic corticosteroid desonide has been clinically tested for the treatment of burns [37]. In another study, conducted on patients subjected to inguinal hair removal with pulsed light, the oil emulsion has induced analgesic and anti-inflammatory effects comparable to those of desonide alone [38]. The wound healing activities of the ethanolic extracts of leaves and bark have been observed in studies on rats [39,40].

The oil is intensively used in the cosmetic industry for lotions, shampoos, creams, and soaps [41]. Its skin care value is mainly due to the abundant presence of linoleic acid, known to be involved in the maintenance of the epidermal layer. The oil is emollient, hydrating, firming, and rejuvenating and has a lenitive effect on irritated skin due to the presence of limonoids. It can be also used as a tonic balm for hair [42].

In a study focused on the oil mechanism of action, the inhibition of glucose-6-phosphate dehydrogenase, which normally promotes fibroblast conversion to adipocytes, has been interpreted as a potential anti-cellulite effect [43]. The depigmenting and anti-wrinkle properties of the oil have also been described [44].

3.5. Adverse Effects

Studies on the systemic toxicity of the oil have been conducted on mice and rats, revealing no disturbing effects, with the exception of slight increments of liver weight and alanine aminotransferase plasma levels [19,45,46].

4. *Eperua falcata* Aubl.

Family Fabaceae, common name: wallaba, parts used: bark. INCI names: Eperua Falcata Bark Extract.

4.1. Features

This is a fast growing tree that can reach a height of 40 m. The leaves are paripinnate with falcate leaflets. The flowers are clustered in reclining inflorescences, and the fruits are falcate pods, hanging from a long peduncle, recalling the blade of a sword. Due to their typical aspect, the fruits are called 'eperu' by natives, meaning 'saber sword' in creole language. The fruits undergo explosive dehiscence at ripening, throwing seeds to distances of tens of meters.

The species is native from Guyana and neighbouring regions of Brazil, Venezuela, and Suriname. It grows on sandy, drained grounds poor in nutrients. In Guyana, it is a main arboreal plant, forming so-called 'wallaba forests', and being exploited as construction timber [47].

4.2. Constituents

The wood contains resinous substances, from which different compounds have been isolated, including the diterpenes eperuic acid and cativic acid (labd-7-en-15-oic acid) [48]. Other compounds isolated from the wood are the flavonol (-)-dihydrokempferol; the glycosylated 3-hyroxyflavanones engeletin, neoengeletin, and astilbin; the flavan-3-ols catechin, epicatechin, and 3-(4-hydroxybenzoyl)-

epicatechin (wallaba epicatechin); the phenolics p-hydroxybenzoic, gallic, and ellagic acids; and proanthocyanidins [49].

4.3. Properties

The bark and wood resin are traditionally used by indigenous populations for toothache, wounds, and articular pain. The wood resin is also an antimycotic and insect repellent [48].

4.4. Dermatologic and Cosmetic Uses

Traditional uses of the plant have inspired scientific research aimed at exploring its anti-inflammatory effects on the skin. These studies have concerned a bark aqueous extract containing proanthocyanidins and flavonoids such as astilbin and engeletin and have been addressed to mechanisms of neuroinflammation mediated by nociceptive terminals. Observed effects include the prevention of pro-inflammatory peptides and cytokines release through the inhibition of the NF-kB pathway in cutaneous nociceptors and keratinocytes. It has also been found that astilbin and engeletin suppress inflammatory processes induced by UV rays on keratinocytes, suggesting that these compounds could play a main role in the anti-inflammatory effects of the extract [50]. Based on these findings, the bark acqueous extract is used in soothing products aimed at alleviating skin redness, rosacea, and chronic micro-inflammatory processes [51].

4.5. Adverse Effects

No harmful or allergic effects have been reported for the plant principles.

5. *Oenocarpus bataua* Mart.

Family: Arecaceae, common name: patawa, parts used: fruit. INCI names: Oenocarpus Bataua Acid, Oenocarpus Bataua Fruit Oil, and Oenocarpus Bataua Seed Oil.

5.1. Features

This is a tree palm, up to 25 m tall, with 10 to 16 apical leaves, each three to seven meters long and having a blade consisting of about 100 segments on each side. The inflorescence is a panicle of small yellow flowers emerging below the leaf crown. The fruit is a red-purple drupe.

The species grows on sandy, acid grounds in wet environments. It is one of the most abundant arboreal plants of the Amazon region and a main source of materials such as wood and leaves. The fruits are also widely used as food, cosmetics, and medicine. Fat extracted by boiling the fruits and collecting the supernatant is used for preparing a milk-like beverage called 'chicha' in Ecuador, 'vino de seje' in Venezuela, and 'vinho de patauá' elsewhere. This drink is an important source of calories and protein in the indigenous diet [52]. Fruits contain 30% lipids (dry weight) that can be converted by cold pressure into yellow-greenish oil, similar to olive oil, which is used as food and cosmetics. Also, in French Guiana and Peru, small local industries produce ice-cream with the palm mesocarp [53,54].

5.2. Constituents

The most abundant oil constituent is the ω-9 oleic acid, followed by palmitic acid. Other major fatty acids include stearic, linoleic, arachidic, pentadecanoic, and α-linolenic acids. The non-saponifiable fraction comprises various sterols such as β-sitosterol, Δ5-avenasterol, stigmasterol, campesterol, campestanol, and cholesterol and, in addition, carotenoids, mainly β-carotene, and tocopherols, mainly α-tocopherol [55,56].

The leaf and root extracts contain hydroxycinnamic acids such as caffeoylquinic and caffeoylshikimic acids and C-glycosyl flavones [57]. Stilbene and its derivative, piceatannol, have been isolated from a methanolic extract of the fruit pulp [58].

5.3. Properties

The fruit pulp is a traditional, multi-purpose remedy against alopecia, cough, bronchitis, tuberculosis, and malaria [56].

5.4. Dermatologic and Cosmetic Uses

People in the Amazon region use the oil as a hair tonic against dandruff, allegedly owing to its antibacterial and antimycotic properties. Moreover, the oil confers strength and brightness to hair [59].

Following its popularity in the regions of origin, the oil has been introduced in the cosmetic industry. Topical applications induce hydrating, thickening, and elasticizing effects, render the skin smooth and silky, and result in a rapid penetration that avoids a greasy feeling. Overall, the oil is an antiage agent and protects the skin from harmful exterior agents, possibly due to its high contents of ω-9 fatty acids, vitamin E (tocopherols), and vitamin A (retinoids). The oil can be also used as a vehicle for the transdermal delivery of lipophilic active principles such as testosterone [60].

5.5. Adverse Effects

No damaging effects are known following topical application of the oil.

6. *Quassia amara* L.

Family: Simaroubaceae, common name: amargo, parts used: wood and bark. INCI names: Quassia Amara Wood, Quassia Amara Wood Extract, Quassia Amara Wood Powder, and Quassin.

6.1. Features

This is a small, evergreen tree, four to six meters high, with a smooth, grayish bark. The leaves are alternate and imparipinnate, with oblong, acuminate leaflets, having winged rachis and reddish veins. The flowers are borne in racemes and have five reddish, spirally twisted petals. The fruit is an aggregate of five obovate-elliptic drupes, attached to a red, fleshy receptacle and containing a single seed each [26].

The species is native to Guiana and northern Brazil but is also found in Venezuela, Colombia, Argentina, Panama, and Mexico. It grows in wet forests at elevations not exceeding 500 m above sea level.

The plant is used as an herbal remedy for the presence of bitter principles. The species *Picrasma excelsa*, known as Jamaican quassia, has properties similar to those of *Q. amara*. Herbal products from either species are frequently indistinctly commercialized under the name of 'quassia'. In addition to herbal ingredients, *Q. amara* is used for the production of flypaper and methylated spirits.

6.2. Constituents

The plant contains seco-triterpene-δ-lactones known as quassinoids, typical of the family Simaroubaceae. The major quassinoids are quassin, neoquassin, 18-hydroxyquassin, simalikalactone D, picrasin, and quassimarin. Most quassinoids have a C-20 structure, but simalikalactone D has a C-25 structure and quassimarin has a C-27 structure [61,62]. The plant also contains indolic alkaloids of the β-carboline family such as 1-vinyl-4,8-dimethoxy-β-carboline, 1-methoxycarbonyl-β-carboline, and 3-methylcanthine-2,6-dione [63].

6.3. Properties

Quassinoids are responsible for most of the plant's biological properties and its strong bitter taste. In southern America, the plant is used as antimalarial and febrifuge substance, as an alternative to quinine root. It is also used as an appetizer and digestive due to its bitter principles and as an insecticide and an antiparasite substance, and it is reported in the Pharmacopeias of different countries [64].

Antimalarial properties have been investigated on animal models, while in vitro studies have revealed cytotoxic activity of simalikalactone D and E on chloroquine-resistant *Plasmodium falciparum* strains [65–67]. In addition, various in vitro and in vivo studies have shown antiviral, insecticidal, anti-inflammatory, gastroprotective, antidiabetic, and antifertility activities of quassinoids [68–73].

β-Carboline alkaloids are a wide family of compounds known as DNA intercalating agents and inhibitors of cyclin-dependent kinases, topoisomerases and monoamino-oxydases. These compounds also interact with benzodiazepine and serotonin receptors, inducing sedative and anxiolytic effects [63].

6.4. Dermatologic and Cosmetic Uses

The infusion is used as a topic treatment for skin parasites like scabies and lice, while clinical investigations have shown prophylactic effects rather than insecticide activity [74,75]. These data are in agreement with studies on the mechanism of action of plant extracts, showing the inhibition of chitin production that prevents louse egg adhesion to hairs [76].

Antifungal and anti-inflammatory properties have been exploited against facial seborrheic dermatitis in a randomized, double-blind study on human patients [77]. Wood and bark extracts are generally used in cosmetics as conditioning and invigorating agents, while antioxidant, anti-inflammatory, and anti-aging properties have also been reported [78]. Different wood extracts have been proposed as either skin collagen stimulants or hair tonics [79,80].

6.5. Adverse Effects

Plant derivatives and quassinoids are generally well tolerated by humans and laboratory animals. Negative effects observed on rat fertility are reversible and not accompanied by damage to other body systems [70]. The plant has been approved as a food supplement by the EU and indicated as Generally Recognized as Safe (GRAS) by the US FDA.

7. Conclusions and Perspectives

We have selected a group of arboreal species that are major elements of the Amazon landscape and culture, being intensively utilized in this region and nearby areas. These plants are valuable sources of oil and other products, but their potentials to generate drugs have been investigated to a limited extent. Even less explored is the possibility of finding applications for dermatologic and cosmetic problems. The use of these products on skin is to a large extent inspired by empiric knowledge, mostly derived from the traditional practices of Amazonian people (Table 1).

Table 1. Reported Effects on Skin and Hair of Phytocomplexes and Active Principles from Amazonian Plants.

Species	Phytocomplexes	Phytochemicals	Effects	Ref.
Attalea speciosa	Fruit extract		Wound healing	[10]
	Seed oil,		Soothing, emollient,	[12]
		Lauric acid, myristic acid	Hydrating, hair volumizing	[11]
Carapa guianensis	Seed oil		Wound healing, insect repellent, anti-psoriasis	[36]
	Seed oil		Emollient, hydrating, firming, depigmenting, rejuvenating	[42]
	Seed oil		Glucose-6-phosphate dehydrogenase inhibition	[43]
	Seed oil/desonide emulsion		Burn healing, analgesic, anti-inflammatory	[37,38]
	Leaf ethanolic extract		Wound healing	[39,40]
		Limonoids	Lenitive	[42]

Table 1. *cont.*

Species	Phytocomplexes	Phytochemicals	Effects	Ref.
Eperua falcata	Bark aqueous extract		Prevention of neuroinflammation through NF-κB inhibition	[50]
	Bark aqueous extract		Lenitive, soothing, anti-rosacea	[51]
		Astilbin, engeletin	Suppression of inflammation induced by UV	[50]
Oenocarpus bataua	Fruit oil		Antibacterial, antimycotic, anti-dandruff	[59]
	Fruit oil		Hydrating, thickening, elasticizing	[60]
Quassia amara	Wood infusion		Anti-parasites, scabies, lice	[74,75]
	Wood and bark extract		Inhibition of louse chitin production	[76]
	Wood and bark extract		Anti-seborrheic dermatitis	[77]
	Wood and bark extract		Antioxidant, anti-inflammatory, anti-aging, collagen stimulant, hair tonic	[78–80]

However, a starting point for the achievement of a scientific basis for skin care and health applications can derive from phytochemical knowledge. Various classes of molecules have been isolated that are potentially interesting in this sense such as the limonoids of *C. guianensis*, the diterpenes and flavanonols of *E. falcata*, and the quassinoids of *Q. amara*. In addition, the abundant lipid fractions of *A. speciosa*, *C. guianensis*, and *O. bataua*, have peculiar compositions, especially in the non-saponifiable fraction, that could provide interesting bioactive molecules. A few pieces of literature have highlighted the importance of these compounds for the development of skin care applications, providing examples for phytosterols [81], limonoids [82], astilbin, and engeletin [83,84]. However, such a complex of knowledge is the tip of an iceberg of hitherto undiscovered bioactivities and constituents in the phytocomplexes of these plant species.

In conclusion, the species examined in this survey are promising sources of bioactive compounds for skin care and health applications. Investigations directed to expand the knowledge about their properties and bioactive agents will give added value to the exploitation of forest resources, possibly helping to orient their utilization toward non-destructive production chains. Following ethical principles and eco-sustainable aims, the Amazon's biodiversity could become a platform for technological research and development of new cosmetic ingredients and products.

Acknowledgments: This work has been financially supported by Helan Cosmesi di Laboratorio S.r.l., Casella, Italy.

Conflicts of Interest: The authors declare no conflict of interest.

References

1. Nikitakis, J.; Lange, B. *International Cosmetic Ingredient Dictionary and Handbook*, 16th ed.; Personal Care Products Council: Washington, DC, USA, 2016.
2. Teixeira, M.A. Babassu—A new approach for an ancient Brazilian biomass. *Biomass Bioenergy* **2008**, *32*, 857–864. [CrossRef]
3. Rufino, M.U.D.; Costa, J.T.D.; da Silva, V.A.; Andrade, L.D.C. Knowledge and use of ouricuri (*Syagrus coronata*) and babacu (*Orbignya phalerata*) in Buique, Pernambuco State, Brazil. *Acta Bot. Bras.* **2008**, *22*, 1141–1149. [CrossRef]
4. Araujo, F.R.; Gonzalez-Perez, S.E.; Lopes, M.A.; Viegas, I.D.M. Ethnobotany of babassu palm (*Attalea speciosa* Mart.) in the Tucurui Lake Protected Areas Mosaic – eastern Amazon. *Acta Bot. Bras.* **2016**, *30*, 193–204. [CrossRef]

5. Santos, D.S.; Silva, I.G.D.; Araújo, B.Q.; Lopes Júnior, C.A.; Monção, N.B.; Citó, A.M.; de Souza, M.H.; Nascimento, M.D.D.; Costa, M.C.P. Extraction and Evaluation of Fatty Acid Compositon of *Orbignya phalerata* Martius Oils (*Arecaceae*) from Maranhao State, Brazil. *J. Braz. Chem. Soc.* **2013**, *24*, 355–362.
6. De Oliveira, A.I.; Mahmoud, T.S.; do Nascimento, G.N.; da Silva, J.F.; Pimenta, R.S.; de Morais, P.B. Chemical Composition and Antimicrobial Potential of Palm Leaf Extracts from Babacu (*Attalea speciosa*), Buriti (*Mauritia flexuosa*), and Macauba (*Acrocomia aculeata*). *Sci. World J.* **2016**, *2016*, 9734181. [CrossRef] [PubMed]
7. Souza, M.H.; Monteiro, C.A.; Figueredo, P.M.; Nascimento, F.R.; Guerra, R.N. Ethnopharmacological use of babassu (*Orbignya phalerata* Mart) in communities of babassu nut breakers in Maranhao, Brazil. *J. Ethnopharmacol.* **2011**, *133*, 1–5. [CrossRef] [PubMed]
8. Balick, M.J. Ethnobotany of palms in the neotropics. *Adv. Econ. Bot.* **1984**, *1*, 9–23.
9. Renno, M.N.; Barbosa, G.M.; Zancan, P.; Veiga, V.F.; Alviano, C.S.; Sola-Penna, M.; Menezes, F.S.; Holandino, C. Crude ethanol extract from babassu (*Orbignya speciosa*): Cytotoxicity on tumoral and non-tumoral cell lines. *An. Acad. Bras. Cienc.* **2008**, *80*, 467–476. [CrossRef] [PubMed]
10. Amorim, E.; Matias, J.E.; Coelho, J.C.; Campos, A.C.; Stahlke, H.J., Jr.; Timi, J.R.; Rocha, L.C.D.A.; Moreira, A.T.R.; Rispoli, D.Z.; Ferreira, L.M. Topic use of aqueous extract of *Orbignya phalerata* (babassu) in rats: Analysis of its healing effect. *Acta Cir. Bras.* **2006**, *21*, 67–76. [CrossRef] [PubMed]
11. Hasegawa, M. Cosmetic Composition. JPS62,192,308, 22 August 1987. Available online: https://worldwide.espacenet.com/publicationDetails/biblio?DB=EPODOC&II=44&ND=3&adjacent= true&locale=en_EP&FT=D&date=19870822&CC=JP&NR=S62192308A&KC=A (accessed on 26 July 2017).
12. Oribuie, K. Cosmetic Composition. JPH04,230,308, 19 August 1992. Available online: https: //worldwide.espacenet.com/publicationDetails/biblio?DB=EPODOC&II=0&ND=3&adjacent=true& locale=en_EP&FT=D&date=19920819&CC=JP&NR=H04230308A&KC=A (accessed on 26 July 2017).
13. Müller, R.; Seidel, K.; Kaczich, A.; Hollenberg, D.; Matzik, I. Skin and Hair Aerosol Foam Preparations Containing an Alkyl Polyglycoside and Vegetable Oil. U.S. Patent 6,045,779, 4 April 2000.
14. Silva, A.P.; Cerqueira, G.S.; Nunes, L.C.; de Freitas, R.M. Effects of an aqueous extract of *Orbignya phalerata* Mart on locomotor activity and motor coordination in mice and as antioxidant in vitro. *Pharmazie* **2012**, *67*, 260–263. [PubMed]
15. Polak, A.M. *Major Timber Trees of Guyana; A Field Guide*; The Tropenbos Foundation: Wageningen, The Netherlands, 1992.
16. Fournier, L.A. Carapa guianenses Aublet. In *Tropical Tree Seed Manual*; Vozzo, J.A., Ed.; USDA Forest Service: Washington, DC, USA, 2003; pp. 360–361.
17. Corrêa, M.P.; de Azeredo Pena, L. *Dicionário das plantas úteis do Brasil e das exóticas cultivadas*; Ministério da Agricultura, Instituto Brasileiro de Desenvolvimento Florestal: Rio de Janeiro, Brazil, 1984.
18. Pinto, G.P. Contribuição ao estudo químico do óleo de Andiroba. *Bol. Tec. Inst. Agron. Norte* **1956**, *31*, 195–206.
19. Milhomem-Paixao, S.S.; Fascineli, M.L.; Roll, M.M.; Longo, J.P.; Azevedo, R.B.; Pieczarka, J.C.; Salgado, H.L.C.; Santos, A.S.; Grisolia, C.K. The lipidome, genotoxicity, hematotoxicity and antioxidant properties of andiroba oil from the Brazilian Amazon. *Genet. Mol. Biol.* **2016**, *39*, 248–256. [CrossRef] [PubMed]
20. Inoue, T.; Matsui, Y.; Kikuchi, Y.; In, Y.; Muraoka, O.; Yamada, T.; Tanaka, R. Carapanolides C-I from the seeds of andiroba (*Carapa guianensis*, Meliaceae). *Fitoterapia* **2014**, *96*, 56–64. [CrossRef] [PubMed]
21. Da Silva, V.P.; Oliveira, R.R.; Figueiredo, M.R. Isolation of limonoids from seeds of *Carapa guianensis* Aublet (Meliaceae) by high-speed countercurrent chromatography. *Phytochem. Anal.* **2009**, *20*, 77–81. [CrossRef] [PubMed]
22. Funasaki, M.; Barroso, H.D.; Fernandes, V.L.A.; Menezes, I.S. Amazon rainforest cosmetics: Chemical approach for quality control. *Quim. Nova* **2016**, *39*, 194–209. [CrossRef]
23. Qi, S.H.; Wu, D.G.; Zhang, S.; Luo, X.D. Constituents of *Carapa guianensis* Aubl. (Meliaceae). *Pharmazie* **2004**, *59*, 488–490. [CrossRef] [PubMed]
24. Andrade, E.H.; Zoghbi, M.D.; Maia, J.G. Volatiles from the leaves and flowers of *Carapa guianensis* Aubl. *J. Essent. Oil Res.* **2001**, *13*, 436–438. [CrossRef]
25. Hammer, M.L.; Johns, E.A. Tapping an Amazonian plethora: Four medicinal plants of Marajo Island, Para (Brazil). *J. Ethnopharmacol.* **1993**, *40*, 53–75. [CrossRef]

26. DeFilipps, R.A.; Maina, S.L.; Crepin, J. *Medicinal Plants of the Guianas (Guyana, Surinam, French Guiana)*; Department of Botany, National Museum of Natural History, Smithsonian Institution: Washington, DC, USA, 2004; Available online: http://botany.si.edu/bdg/medicinal/Medicinal_plants_master.pdf (accessed on 26 July 2017).

27. Penido, C.; Conte, F.P.; Chagas, M.S.; Rodrigues, C.A.; Pereira, J.F.; Henriques, M.G. Antiinflammatory effects of natural tetranortriterpenoids isolated from *Carapa guianensis* Aublet on zymosan-induced arthritis in mice. *Inflamm. Res.* **2006**, *55*, 457–464. [CrossRef] [PubMed]

28. Penido, C.; Costa, K.A.; Pennaforte, R.J.; Costa, M.F.; Pereira, J.F.; Siani, A.C.; Henriques, M.G. Anti-allergic effects of natural tetranortriterpenoids isolated from *Carapa guianensis* Aublet on allergen-induced vascular permeability and hyperalgesia. *Inflamm. Res.* **2005**, *54*, 295–303. [CrossRef] [PubMed]

29. Henriques, M.; Penido, C. The therapeutic properties of *Carapa guianensis*. *Curr. Pharm. Des.* **2014**, *20*, 850–856. [CrossRef] [PubMed]

30. Sarria, A.L.; Soares, M.S.; Matos, A.P.; Fernandes, J.B.; Vieira, P.C.; da Silva, M.F. Effect of triterpenoids and limonoids isolated from *Cabralea canjerana* and *Carapa guianensis* (Meliaceae) against Spodoptera frugiperda (J. E. Smith). *Z. Naturforsch. C* **2011**, *66*, 245–250. [CrossRef] [PubMed]

31. Silva, O.S.; Romao, P.R.; Blazius, R.D.; Prohiro, J.S. The use of andiroba *Carapa guianensis* as larvicide against Aedes albopictus. *J. Am. Mosq. Control Assoc.* **2004**, *20*, 456–457. [PubMed]

32. Omar, S.; Zhang, J.; MacKinnon, S.; Leaman, D.; Durst, T.; Philogene, B.J.; Arnason, J.T.; Sanchez-Vindas, P.E.; Poveda, L.; Tamez, P.A.; et al. Traditionally-used antimalarials from the Meliaceae. *Curr. Top. Med. Chem.* **2003**, *3*, 133–139. [CrossRef] [PubMed]

33. Kamath, S.G.; Chen, N.; Xiong, Y.; Wenham, R.; Apte, S.; Humphrey, M.; Cragun, J.; Lancaster, J.M. Gedunin, a novel natural substance, inhibits ovarian cancer cell proliferation. *Int. J. Gynecol. Cancer* **2009**, *19*, 1564–1569. [CrossRef] [PubMed]

34. Miranda Junior, R.N.; Dolabela, M.F.; da Silva, M.N.; Povoa, M.M.; Maia, J.G. Antiplasmodial activity of the andiroba (*Carapa guianensis* Aubl., Meliaceae) oil and its limonoid-rich fraction. *J. Ethnopharmacol.* **2012**, *142*, 679–683. [CrossRef] [PubMed]

35. Henriques, M.G.; Monteiro, C.P.; Siani, A.C.; de Fatima, G. J.; de Souza, R.M.F.; Sampaio, A.L.F.; Rosas, E.C.; de Lima, L.A.; Pennaforte, R.J. Pharmaceutical Compositions from *Carapa guianensis*. US2012148695, 14 June 2012. Available online: https://worldwide.espacenet.com/publicationDetails/biblio?DB=EPODOC&II=0&ND=3&adjacent=true&locale=en_EP&FT=D&date=20120614&CC=US&NR=2012148695A1&KC=A1 (accessed on 26 July 2017).

36. Pesso, J. Pediculicide Compositions. Patent EP2,303,408, 14 December 2011.

37. Cela, E.V.S.S.; Rocha, M.B.; Chia, C.Y.; Alves, C.F. Treatment of firstdegree burns with andiroba oil emulsion: A prospective, comparative, doubleblind study. *Surg. Cosmet. Dermatol.* **2014**, *6*, 44–49.

38. Cela, E.V.S.S.; da Rocha, M.B.; Gomes, T.M.; Chia, C.Y.; Alves, C.F. Clinical evaluation of the effec tiveness of andiroba oil in burns caused by hair removal with intense pulsed light: A prospective, comparative and double-blind study. *Surg. Cosmet. Dermatol.* **2012**, *4*, 248–251.

39. Nayak, B.S.; Kanhai, J.; Milne, D.M.; Pinto Pereira, L.; Swanston, W.H. Experimental evaluation of ethanolic extract of *Carapa guianensis* L. Leaf for its wound healing activity using three wound models. *Evid. Based Complement. Alternat. Med.* **2011**, *2011*, 419612. [CrossRef] [PubMed]

40. Nayak, B.S.; Kanhai, J.; Milne, D.M.; Swanston, W.H.; Mayers, S.; Eversley, M.; Rao, A.C. Investigation of the wound healing activity of *Carapa guianensis* L. (Meliaceae) bark extract in rats using excision, incision, and dead space wound models. *J. Med. Food* **2010**, *13*, 1141–1146. [CrossRef] [PubMed]

41. Ferreira, M.R.; Santiago, R.R.; de Souza, T.P.; Egito, E.S.; Oliveira, E.E.; Soares, L.A. Development and evaluation of emulsions from *Carapa guianensis* (Andiroba) oil. *AAPS PharmSciTech* **2010**, *11*, 1383–1390. [CrossRef] [PubMed]

42. Franquilino, E. Ativos amazônicos. *Cosmet. Toilet.* **2006**, *18*, 18–53.

43. Rouillard, F.; Crepin, J.; Saintigny, G. Cosmetic or Pharmaceutical Composition Containing an Andiroba Extract. U.S. Patent 5,958,421, 28 September 1999.

44. Park, D.H.; Kim, S.B.; Lee, J.S. Cosmetic Compositions Comprising Andiroba Oils. KR20100013095, 9 February 2010. Available online: https://worldwide.espacenet.com/publicationDetails/biblio?DB=EPODOC&II=0&ND=3&adjacent=true&locale=en_EP&FT=D&date=20100209&CC=KR&NR=20100013095A&KC=A (accessed on 26 July 2017).

45. Costa-Silva, J.H.; Lima, C.R.; Silva, E.J.; Araujo, A.V.; Fraga, M.C.; Ribeiro, E.R.A.; Arruda, A.C.; Lafayette, S.S.L.; Wanderley, A.G. Acute and subacute toxicity of the *Carapa guianensis* Aublet (Meliaceae) seed oil. *J. Ethnopharmacol.* **2008**, *116*, 495–500. [CrossRef] [PubMed]

46. Costa-Silva, J.H.; Lyra, M.M.; Lima, C.R.; Arruda, V.M.; Araujo, A.V.; e Ribeiro, A.R.; Arruda, A.C.; Fraga, M.C.C.A.; Lafayette, S.S.L.; Wanderley, A.G. A toxicological evaluation of the effect of *Carapa guianensis* Aublet on pregnancy in Wistar rats. *J. Ethnopharmacol.* **2007**, *112*, 122–126. [CrossRef] [PubMed]

47. Vink, A.T. *Surinam Timbers: A Summary of Available Information with Brief Descriptions of the Main Species of Surinam*, 3rd ed.; Surinam Forest Service: Paramaribo, Suriname, 1965.

48. Amusant, N.; Moretti, C.; Richard, B.; Prost, E.; Nuzillard, J.M.; Thevenon, M.F. Chemical compounds from *Eperua falcata* and *Eperua grandiflora* heartwood and their biological activities against wood destroying fungus (*Coriolus versicolor*). *Holz. Roh. Werkst.* **2007**, *65*, 23–28. [CrossRef]

49. Royer, M.; Stien, D.; Beauchene, J.; Herbette, G.; McLean, J.P.; Thibaut, A.; Thibaut, B. Extractives of the tropical wood wallaba (*Eperua falcata* Aubl.) as natural anti-swelling agents. *Holzforschung* **2010**, *64*, 211–215. [CrossRef]

50. Henry, F.; Moser, P.; Danoux, L.; Contet-Audonneau, J.L.; Pauly, G. Preparations Containing an Extract of Eperua falcata and/or Constituents of the Latter. U.S. Patent 20070003510, 4 January 2007.

51. Baldwin, H.; Berson, D.; Vitale, M.; Yatskayer, M.; Chen, N.; Oresajo, C. Clinical effects of a novel topical composition on persistent redness observed in patients who had been successfully treated with topical or oral therapy for papulopustular rosacea. *J. Drugs Dermatol.* **2014**, *13*, 326–331. [PubMed]

52. Balick, M.J.; Gershoff, S.N. Nutritional evaluation of the Jessenia bataua palm: Source of high quality protein and oil from Tropical America. *Econ. Bot.* **1981**, *35*, 261–271. [CrossRef]

53. De Almeida, M.E. Sôbre a semelhança dos óleos de patauá e de oliva, sua diferenciação. *Rev. Inst. Adolfo Lutz* **1953**, *13*, 57–65. [PubMed]

54. Galeano, G. *Las Palmas de la región del Araracuara*, 2nd ed.; Topembos, Universidad Nacional: Bogotá, Colombia, 1992.

55. Darnet, S.H.; da Silva, L.H.M.; Rodrigues, A.M.D.; Lins, R.T. Nutritional composition, fatty acid and tocopherol contents of buriti (*Mauritia flexuosa*) and patawa (*Oenocarpus bataua*) fruit pulp from the Amazon region. *Cienc. Tecnol. Aliment.* **2011**, *31*, 488–491. [CrossRef]

56. Montufar, R.; Laffargue, A.; Pintaud, J.C.; Hamon, S.; Avallone, S.; Dussert, S. *Oenocarpus bataua* Mart. (*Arecaceae*): Rediscovering a Source of High Oleic Vegetable Oil from Amazonia. *J. Am. Oil Chem. Soc.* **2010**, *87*, 167–172. [CrossRef]

57. Leba, L.J.; Brunschwig, C.; Saout, M.; Martial, K.; Bereau, D.; Robinson, J.C. *Oenocarpus bacaba* and *Oenocarpus bataua* Leaflets and Roots: A New Source of Antioxidant Compounds. *Int. J. Mol. Sci.* **2016**, *17*, 1014. [CrossRef] [PubMed]

58. Hidalgo, P.S.P.; Nunomura, R.D.C.S.; Nunomura, S.M. Amazon oilseeds: Chemistry and antioxidant activity of patawa (*Oenocarpus bataua* Mart.). *Rev. Virtual Quim.* **2016**, *8*, 130–140. [CrossRef]

59. Shanley, P.; Medina, G. *Frutíferas e plantas úteis na vida amazônica*; CIFOR and IMAZON: Belem, Brazil, 2005.

60. Tsu-I, C.W.; Biundo, B.V. Testosterone Booster Transdermal Compositions. U.S. Patent 20150065426, 5 March 2015.

61. Guo, Z.; Vangapandu, S.; Sindelar, R.W.; Walker, L.A.; Sindelar, R.D. Biologically active quassinoids and their chemistry: Potential leads for drug design. *Curr. Med. Chem.* **2005**, *12*, 173–190. [CrossRef] [PubMed]

62. Houel, E.; Bertani, S.; Bourdy, G.; Deharo, E.; Jullian, V.; Valentin, A.; Chevalley, S.; Stien, D. Quassinoid constituents of *Quassia amara* L. leaf herbal tea. Impact on its antimalarial activity and cytotoxicity. *J. Ethnopharmacol.* **2009**, *126*, 114–118. [CrossRef] [PubMed]

63. Cao, R.; Peng, W.; Wang, Z.; Xu, A. Beta-Carboline alkaloids: Biochemical and pharmacological functions. *Curr. Med. Chem.* **2007**, *14*, 479–500. [CrossRef] [PubMed]

64. Saez, J.A.L.; Soto, J.P. Ethnopharmacology and biological activity of *Quassia amara* (Simaroubaceae): State of the art. *Boletin Latinoam. Caribe Plantas Med. Aromat.* **2008**, *7*, 234–246.

65. Kirby, G.C.; O'Neill, M.J.; Phillipson, J.D.; Warhurst, D.C. In vitro studies on the mode of action of quassinoids with activity against chloroquine-resistant Plasmodium falciparum. *Biochem. Pharmacol.* **1989**, *38*, 4367–4374. [CrossRef]

66. Cachet, N.; Hoakwie, F.; Bertani, S.; Bourdy, G.; Deharo, E.; Stien, D.; Houel, E.; Gornitzka, H.; Fillaux, J.; Chevalley, S.; et al. Antimalarial activity of simalikalactone E, a new quassinoid from *Quassia amara* L. (Simaroubaceae). *Antimicrob. Agents Chemother.* **2009**, *53*, 4393–4398. [CrossRef] [PubMed]

67. Bertani, S.; Houel, E.; Stien, D.; Chevolot, L.; Jullian, V.; Garavito, G.; Bourdy, G.; Deharo, E. Simalikalactone D is responsible for the antimalarial properties of an Amazonian traditional remedy made with *Quassia amara* L. (Simaroubaceae). *J. Ethnopharmacol.* **2006**, *108*, 155–157. [CrossRef] [PubMed]

68. Toma, W.; Gracioso, J.S.; Hiruma-Lima, C.A.; Andrade, F.D.; Vilegas, W.; Souza Brito, A.R. Evaluation of the analgesic and antiedematogenic activities of *Quassia amara* bark extract. *J. Ethnopharmacol.* **2003**, *85*, 19–23. [CrossRef]

69. Psota, V.; Ourednickova, J.; Falta, V. Control of Hoplocampa testudinea using the extract from *Quassia amara* in organic apple growing. *Hortic. Sci.* **2010**, *37*, 139–144.

70. Parveen, S.; Das, S.; Kundra, C.P.; Pereira, B.M. A comprehensive evaluation of the reproductive toxicity of *Quassia amara* in male rats. *Reprod. Toxicol.* **2003**, *17*, 45–50. [CrossRef]

71. Husain, G.M.; Singh, P.N.; Singh, R.K.; Kumar, V. Antidiabetic activity of standardized extract of *Quassia amara* in nicotinamide-streptozotocin-induced diabetic rats. *Phytother. Res.* **2011**, *25*, 1806–1812. [CrossRef] [PubMed]

72. Garcia-Barrantes, P.M.; Badilla, B. Anti-ulcerogenic properties of *Quassia amara* L. (Simaroubaceae) standardized extracts in rodent models. *J. Ethnopharmacol.* **2011**, *134*, 904–910. [CrossRef] [PubMed]

73. Apers, S.; Cimanga, K.; Vanden Berghe, D.; Van Meenen, E.; Longanga, A.O.; Foriers, A.; Vlietinck, A.; Pieters, L. Antiviral activity of simalikalactone D, a quassinoid from *Quassia africana*. *Planta Med.* **2002**, *68*, 20–24. [CrossRef] [PubMed]

74. Rossini, C.; Castillo, L.; González, A. Plant extracts and their components as potential control agents against human head lice. *Phytochem. Rev.* **2008**, *7*, 51–63. [CrossRef]

75. Ninci, M.E. Prophylaxis and treatment of pediculosis with *Quassia amarga*. *Rev. Fac. Cien. Med. Univ. Nac. Cordoba* **1991**, *49*, 27–31. [PubMed]

76. Alcalde, M.T.; Del Pozo, A. Vinagre de quassia como tratamiento cosmético natural contra los piojos. *Offarm* **2007**, *26*, 132–133.

77. Diehl, C.; Ferrari, A. Efficacy of topical 4% *Quassia amara* gel in facial seborrheic dermatitis:a randomized, double-blind, comparative study. *J. Drugs Dermatol.* **2013**, *12*, 312–315. [PubMed]

78. Sogabe, T.; Shiyuu, E.; Kishida, N.; Oto, N. Skin Cosmetic. JP2003081850, 19 March 2003. Available online: https://worldwide.espacenet.com/publicationDetails/biblio?DB=EPODOC&II=0&ND=3&adjacent=true&locale=en_EP&FT=D&date=20030319&CC=JP&NR=2003081850A&KC=A (accessed on 26 July 2017).

79. Cenizo, V.; Andre, V. Cosmetic Method for Increasing Collagen Expression in Skin Comprising Topically Applying an Extract of *Quassia amara*. U.S. Patent 20150056310, 26 February 2015.

80. Fuhrmann, G.; Giesen, M. Cosmetic, Non-Therapeutic Use of Extract of *Quassia amara* e.g. for Improving Structure of Keratinous Fibers for Improving Combability Hair, Improving Grip of Hair, Increasing Gloss of Hair, and Increasing Hair Keratin Expression. DE Patent 102013213027, 10 April 2014.

81. Puglia, C.; Bonina, F. In vivo spectrophotometric evaluation of skin barrier recovery after topical application of soybean phytosterols. *J. Cosmet. Sci.* **2008**, *59*, 217–224. [PubMed]

82. Akihisa, T.; Noto, T.; Takahashi, A.; Fujita, Y.; Banno, N.; Tokuda, H.; Koike, K.; Suzuki, T.; Yasukawa, K.; Kimura, Y. Melanogenesis inhibitory, anti-inflammatory, and chemopreventive effects of limonoids from the seeds of *Azadirachta* indicia A. Juss. (neem). *J. Oleo Sci.* **2009**, *58*, 581–594. [CrossRef] [PubMed]

83. Huang, H.; Cheng, Z.; Shi, H.; Xin, W.; Wang, T.T.; Yu, L.L. Isolation and characterization of two flavonoids, engeletin and astilbin, from the leaves of *Engelhardia roxburghiana* and their potential anti-inflammatory properties. *J. Agric. Food Chem.* **2011**, *59*, 4562–4569. [CrossRef] [PubMed]

84. Di, T.T.; Ruan, Z.T.; Zhao, J.X.; Wang, Y.; Liu, X.; Wang, Y.; Li, P. Astilbin inhibits Th17 cell differentiation and ameliorates imiquimod-induced psoriasis-like skin lesions in BALB/c mice via Jak3/Stat3 signaling pathway. *Int. Immunopharmacol.* **2016**, *32*, 32–38. [CrossRef] [PubMed]

cosmetics

MDPI

Article

Evaluation of the Effect of Plant Mixture Ethanol Extracts Containing *Biota orientalis* L. Extract on Suppression of Sebum in Cultured Sebocytes and on Stimulation of Growth of Keratinocytes Co-cultured with Hair Papilla Cells

Haifeng Zeng [1], Lihao Gu [1] and Kazuhisa Maeda [1,2,*]

[1] Bionics Program, Tokyo University of Technology Graduate School, 1404-1 Katakuramachi, Hachioji City, Tokyo 192-0982, Japan; SteveZeng@hotmail.com (H.Z.); chaos19901008@gmail.com (L.G.)
[2] School of Bioscience and Biotechnology, Tokyo University of Technology, 1404-1 Katakuramachi, Hachioji City, Tokyo 192-0982, Japan
* Correspondence: kmaeda@stf.teu.ac.jp; Tel./Fax: +81-42-6372442

Received: 22 July 2017; Accepted: 11 August 2017; Published: 14 August 2017

Abstract: *Biota orientalis* L. leaf extract (BOLE) is used medically to improve strength and arrest hemorrhage. In China, BOLE has been used in traditional medicine for its antibacterial properties and for hair restoration. In this study, we investigated the mechanism of hair restoration by BOLE from the point of view of the sebum suppressant effect and hair loss prevention. BOLE at 25 or 50 µg/mL final concentrations, a hair growth plant ethanol extract (HGPEE), and a hair growth plant water extract (HGPWE) (the latter two each containing BOLE and other plant compounds), were used to study: (1) the sebum suppressant effect in sebocytes from normal golden hamster ear pinna origin; (2) the effect on the growth of human fetal epidermal keratinocytes; and (3) the effect on gene expression related to hair growth stimulation, with (2) and (3) studied in human fetal epidermal keratinocytes and hair papilla cells. BOLE had a sebum depletion effect in cultured sebocytes; moreover, the amounts of mRNA of the hair growth factors, KGF, VEGF, and G3PDH analyzed by real-time polymerase chain reaction in human hair papilla cells were increased by HGPEE. The amount of mRNA of Wnt10b in cultured epidermal keratinocytes was increased by the addition of BOLE, and the growth of the cultured epidermal keratinocytes was promoted by HGPEE in a two-layer culture system of hair papilla cells and epidermal keratinocytes. HGPEE had a hair growth promotion/hair restoration effect and a sebum suppression effect. Hair restorers containing HGPEE may be useful for stimulating hair growth and suppressing excess scalp sebum in males and females.

Keywords: *Biota orientalis*; *Platycladus orientalis*; *Thuja orientalis*; sebocytes; scalp sebum; hair growth

1. Introduction

Sometimes the skin excretes oil from pores in excessive amounts. The scalp has a higher quantity of sebum than the face, with excessive oil also excreted from pores. Excessive lipid content on the top of the head (more than on the temporal and occipital regions) contributes to hair loss. The scalp's sebum, containing triglycerides, fatty acids, squalene and wax esters, are produced by the sebaceous gland [1]. Other lipids on the scalp's surface, such as cholesterol esters and cholesterol, mainly originate from epidermal cells [1]. There are more wax esters and fatty acids in the scalp than those on facial skin. Triglycerides are broken down by lipase, which results in a fatty acid. Unsaturated fatty acids are easily oxidized in the air or by sun exposure and are converted to peroxides after a few hours, which causes hair loss and thin hair. Therefore, to maintain healthy hair, it is necessary to avoid excessive scalp lipids.

Isolation cultures of human sebaceous gland cells (sebocytes) are difficult, so has been performed using hamster sebocytes in biosynthetic research for the study of lipid regulation in sebaceous glands [2]. As hamster sebocytes accumulate lipid droplets through cell differentiation, these cells have been widely used for lipid metabolism research, as well as a cell model system of lipid metabolism using normal cells. Moreover, it is known that the lipids of hamster sebocytes increase with the addition of male hormones (dihydrotestosterone), and the cell characteristics are similar to those of human sebocytes [3]. Therefore, hamster sebocytes have been considered to be useful healthy cells alternatives to human sebocytes in sebaceous gland research.

Hair is an appendage of the skin that is peculiar to mammals. The hair follicle is formed by the interaction between the epidermis and dermis, and hair grows and falls out repeatedly in the hair cycle. In recent years, it has been demonstrated that wingless-type MMTV integration sites (Wnt) are important for hair growth [4]. Wnt10b, which develops in a matrix in the early stage of secondary hair germ, induces formation of secondary hair germ and expression of Sonic hedgehog (Shh); Shh then induces Wnt5a in hair papilla cells [5]. It has also been reported that mature hair follicles are not formed as the expression of Wnt5a in hair papilla cells was not observed in Shh knockout mice [6]. Wnt10b is an important signal protein between cells in the generation of hair follicles and promotes the cell growth of hair follicles, as shown by a study where Wnt10b promoted hair shaft growth in the organ cultures of mice mustache hair follicles [7]. The number of formed hair follicles was shown to decrease to one-third of those observed in lymphoid enhancer-binding factor 1 (Lef 1) (the transcription factor of the Wnt pathway) knockout mice [8]. Moreover, it has been reported that hair in the anagen phase also required hair growth factors such as vascular endothelial growth factor (VEGF) and keratinocyte growth factor (KGF) produced by papilla cells. VEGF has decreased expression in hair tissue in male pattern baldness (androgenetic alopecia). VEGF increases the papilla cells in an autocrine manner [9] and the amount of VEGF expression decreases from the anagen phase to the catagen phase [10,11]. When KGF is injected into nude mice, hair growth is promoted, and the hair follicles develop KGF receptors [12,13].

Biota orientalis (*B. orientalis*) is an evergreen needle-leaf tree of the *Platycladus orientalis* genus of the cypress family that originated in China. In Japan, *P. orientalis* can be planted broadly from southern Hokkaido to Kyushu. The leaf is used medicinally for strength nourishment and hemorrhage arrest [14]. In Japan, a folk medicine extract made by dipping the leaf of dried *B. orientalis* in white distilled liquor has been used to prevent hair loss and promote hair restoration instead of modern hair tonics. Moreover, in China, *B. orientalis* has been used for its antibacterial properties and for hair restoration as a tradition medical treatment [14]. It has been reported that a hot water extract of the leaf of *Thuja orientalis* had a hair restoration effect in mice [15]. Furthermore, the seed of *Thuja occidentalis*, which is related to *B. orientalis*, has an inhibitory action of 5α-reductase, and the active ingredients have been reported as flavonoid and diterpene [16].

In this study, the mechanism of hair restoration caused by an ethanol extract of *B. orientalis* (BOLE) was studied by assessing its sebum-suppressant and hair loss-preventative effects. The results demonstrated that BOLE inhibited lipid production in cultured sebocytes. It was also proven that an ethanol extract of a plant extract mixture that contained *B. orientalis* leaves inhibited lipid production in cultured sebocytes, increasing the amounts of mRNA in VEGF and KGF, which are hair growth factors in hair papilla cells. In addition, an increase in the amount of mRNA of Wnt10b in cultured epidermal keratinocytes was observed. Furthermore, the growth of epidermal keratinocytes was promoted by a two-layer culture system of hair papilla cells and epidermal keratinocytes.

2. Materials and Methods

2.1. Materials

The following plant extracts were obtained from Yunnan Baiyao Industry Co. Ltd. (Yunnan, China): ethanol extract of *B. orientalis* L. leaf (Cupressaceae) (BOLE); hair growth plant mixture ethanol extract (HGPEE) consisting of *B. orientalis* L. (Cupressaceae) leaves, *Eclipta thermalis* (Compositae), *Sophora angustifolia* (Leguminosae) root, *Cnidium monnieri* (Umbelliferae) fruit, *Ligusticum chuanxiong* (Apiaceae) rhizome, and *Panax notoginseng* Burk. (Araliaceae); hair growth plant mixture water extract (HGPWE) consisting of *B. orientalis* L. (Cupressaceae) leaves, *E. thermalis* (Compositae) grass, *S. angustifolia* (Leguminosae), *Cnidium monnieri* (Umbelliferae) fruit, *L. chuanxiong* (Apiaceae) rhizome, and *P. notoginseng* Burk. (Araliaceae). Minoxidil was purchased from Sigma-Aldrich Corp. (St. Louis, MO, USA).

2.2. Preparation of Samples

BOLE, HGPEE, and HGPWE were dissolved in dimethyl sulfoxide (DMSO):phosphate-buffered saline (PBS) (1:1) and each diluted to concentrations of 10 mg/mL. These samples were prepared as undiluted solution and as 2-fold, 4-fold, and 8-fold dilutions in DMSO:PBS (1:1) to give 10,000, 5000, 2500, and 1250 μg/mL concentrations, respectively. Minoxidil was prepared at concentrations of 2000, 1000, and 500 μg/mL. DMSO:PBS (1:1) was used as the control (solvent).

2.3. Measurement of Lipids in Sebocytes

Sebocytes obtained from normal golden hamster ear pinna were seeded into a 24-well plate (AGC Techno Glass Co., Ltd., Shizuoka, Japan) at a density of 5.0×10^4 cells/well. The sebocytes were cultured for several days in Dulbecco's Modified Eagle's Medium (DMEM):Ham's F12 (1:1) growth culture medium that contained 8% bovine serum, 2% human serum, and 8% 10 ng/mL epidermal growth factor (EGF). Next, the medium was exchanged for a differentiation DMEM:Ham's F12 (1:1) culture medium that contained 8% bovine serum, 2% human serum, and 10 μg/mL insulin (in addition to the BOLE, HGPEE, and HGPWE examination samples), and culturing was continued for 1 week. BOLE was examined at 25 μg/mL and 50 μg/mL final concentrations, and HGPEE and HGPWE were examined at 50 μg/mL final concentrations. In addition, minoxidil was examined at 20 μg/mL final concentration as a positive control. Then, 50 μL of Cell Counting Kit-8 (Dojindo Laboratories, Kumamoto, Japan) was added to each well and incubated at 37 °C for 2 h. The supernatant was measured at a wavelength of 450 nm using a microplate reader (Multi-Detection Microplate POWERSAN HT; BioTek Instruments Inc., Winooski, VT, USA) and the number of viable cells was measured. Furthermore, the cell was fixed at room temperature for 10 min with 4% paraformaldehyde solution after washing each well with PBS. Each well was washed with PBS, then replaced with isopropanol 60% after washing for 1 min, the isopropanol was discarded, 300 μL of filtered 0.3% oil red O stain solution (60% isopropanol) was added, and it was dyed at room temperature for 30 min. Washing with 60% isopropanol was performed once, followed by washing with PBS twice, and observation in PBS under a microscope. The PBS was discarded, and the lipids that were previously dyed were extracted by adding of 400 μL of 100% isopropanol. The lipids were then measured at a wavelength of 520 nm 30 min later, and the amount of lipids was measured.

The following formula was used to calculate the amount of lipids per cell.

$$\text{Amount of lipids per cell} = \text{absorbance B}/\text{absorbance A} \tag{1}$$

where absorbance A (450 nm) reflected the number of cells and absorbance B (520 nm) reflects the amount of lipids.

2.4. Effect on Growth of Human Fetal Epidermal Keratinocytes in a Two-Layer Culture System of Human Hair Papilla Cells and Human Epidermal Keratinocytes

Human fetal epidermal keratinocytes were cultured in human epidermal keratinocyte growth medium. The cells were prepared at 60,000 cells/well, and 500 μL/well was seeded into a 24-well plate (Falcon; Thermo Fisher Scientific Inc., Waltham, MA, USA) and cultured for 1 day in the epidermal keratinocyte basic culture medium. Human hair papilla cells (Toyobo Co., Ltd., Osaka, Japan) were seeded into 75 cm^2 flasks (3,000,000 cells/flask) and cultured for 1 day in a 10-mL human papilla cell growth culture medium. Next, a 100,000 cells/well, and 100 μL/well were placed in the cell culture insert for a 24-well plate (Falcon) and cultured for 1 day in the human papilla cell basic culture medium. The cells were cultured in human papilla cell basic culture medium containing the control, BOLE (at final concentrations of 12.5, 25, and 50 μg/mL), HGPEE (at final concentrations of 12.5, 25, and 50 μg/mL), HGPWE (at final concentrations of 12.5, 25, and 50 μg/mL), and minoxidil (at final concentrations of 5, 10, and 20 μg/mL) in the cell culture insert for 5 days. The cell culture insert was removed, a 50 μL/well was added to a Cell Counting Kit-8 (Dojindo Laboratories) to the lower layer, culturing proceeded for 2 h, and the number of the human epidermal keratinocytes was measured using a microplate reader (450 nm). Four replicate experiments were performed.

2.5. Influence on Gene Expression Related to Hair Growth/Restoration

Human fetal epidermal keratinocytes (Toyobo) were seeded into a 35-mm dish (300,000 cells/dish) and cultured for 1 day in 1 mL of human epidermal keratinocyte growth medium. An RNA extraction kit (RNeasy Protect Mini Kit; QIAGEN K.K., Tokyo, Japan) was used after culture in the basic culture medium of the control, HGPEE (final concentration of 50 μg/mL), HGPWE (final concentration of 50 μg/mL), and minoxidil (final concentration of 20 μg/mL) for 3 days. The amounts of mRNA of Wnt10b, Lef1, Shh, and G3PDH were analyzed using real-time polymerase chain reaction (RT-PCR) (ABI PRISM 7900HT apparatus) and a TaKaRa One Step SYBR PrimeScript RT-PCR Kit II (Takara Bio Inc., Shiga, Japan). Primers of Wnt10b, Lef1, Shh, and G3PDH were purchased from QIAGEN.

Human hair papilla cells (Toyobo) were seeded into a 35-mm dish (300,000 cells/dish) and cultured for 1 day in 1 mL of human papilla cell growth culture medium. After culture in the basic culture medium of the control, HGPEE (final concentrations of 12.5 and 25 μg/mL), HGPWE (final concentration of 50 μg/mL), and minoxidil (final concentration of 20 μg/mL) for 3 days, an RNA extraction kit (RNeasy Protect Mini Kit, QIAGEN K.K., Tokyo, Japan) was used to perform RT-PCR. The amounts of mRNA of KGF, VEGF, and G3PDH were analyzed using RT-PCR (ABI PRISM 7900HT apparatus) and a TaKaRa One Step SYBR PrimeScript RT-PCR Kit II (Takara Bio Inc.). Primers of KGF, VEGF, and G3PDH were purchased from Qiagen.

2.6. Statistical Analysis

Tests for statistical significance between unpaired groups with normal distribution were performed by using Student's *t*-test in the case of homogeneity of variance and with Welch's *t*-test in the case of unequal variance, and $p < 0.05$ indicated statistical significance.

3. Results

3.1. Effects of Biota Orientalis L. Leaf Extract (BOLE), Hair Growth Plant Ethanol Extract (HGPEE), Hair Growth Plant Ethanol Extract (HGPEE), and Minoxidil on the Proliferation of Sebocytes and Amount of Lipid Droplets

We investigated the mechanism of hair restoration by BOLE on the basis of the sebum suppressant effect. Cell proliferation in the differentiation culture medium was significantly inhibited to three-fifths that of the non-differentiation culture medium (Figure 1). The amount of lipids per cell increased significantly by 2.3-fold and the intracellular lipid droplets also increased significantly by 7.4-fold (Figure 1). There were no significant differences in cell proliferation among the control and BOLE (25, 50 μg/mL), HGPEE (50 μg/mL), and HGPWE (50 μg/mL) in the differentiation culture medium; however, cell proliferation was promoted by minoxidil (20 μg/mL) relative to that of the control (Figure 1).

The size of the intracellular lipid droplet became significantly small after treatment with BOLE (25, 50 μg/mL) and HGPEE (50 μg/mL) relative to the size after treatment with the control, and the size tended to decrease after treatment with HGPWE (50 μg/mL) in the differentiated culture medium. No significant difference in size was seen between minoxidil (20 μg/mL) and the control in the differentiated culture medium (Figure 1).

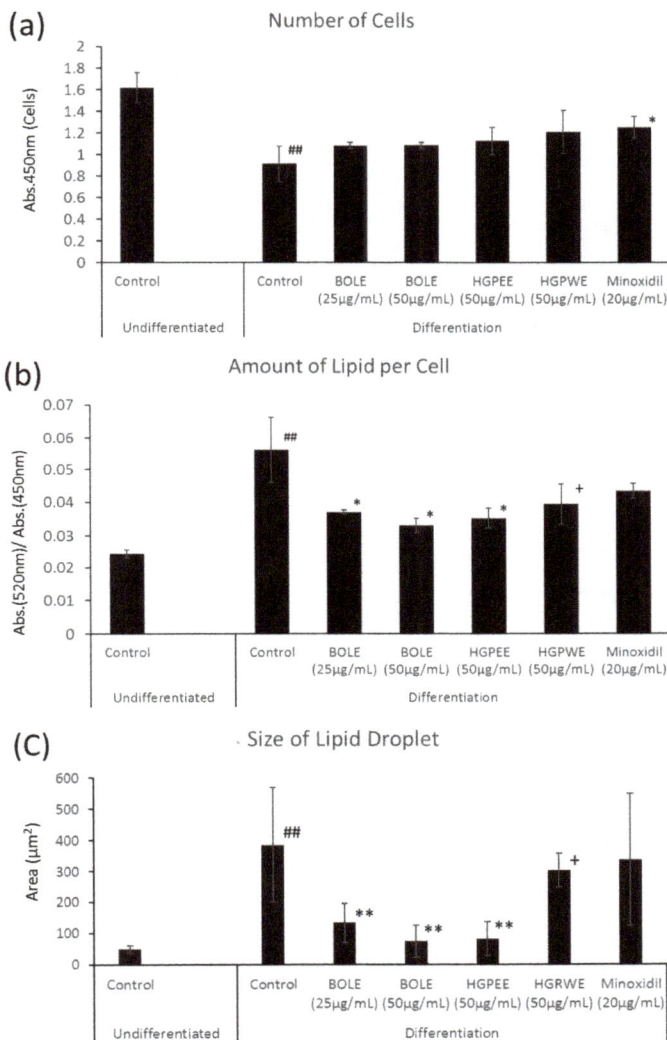

Figure 1. Effects of *Biota orientalis* L. leaf extract (BOLE), hair growth plant ethanol extract (HGPEE), hair growth plant ethanol extract (HGPEE), and minoxidil on the number of hamster sebocytes (**a**); amount of lipids per cell (**b**); and size of lipid droplets (**c**). n = 3, mean ± S.D. $^{\#\#}$ $p < 0.01$ vs. Control (Undifferentiated); * $p < 0.05$ vs. Control (Differentiation); ** $p < 0.01$ vs. Control (Differentiation); $^{+}$ $p < 0.1$ vs. Control (Differentiation).

The shape of the sebocytes dyed with oil red O observed under microscope is shown in Figure 2. The cells were undifferentiated sebocytes, differentiated sebocytes, differentiated sebocytes + BOLE (50 μg/mL), differentiated sebocytes + HGPEE (50 μg/mL), differentiated sebocytes + HGPWE (50 μg/mL), and differentiated sebocytes + minoxidil (20 μg/mL).

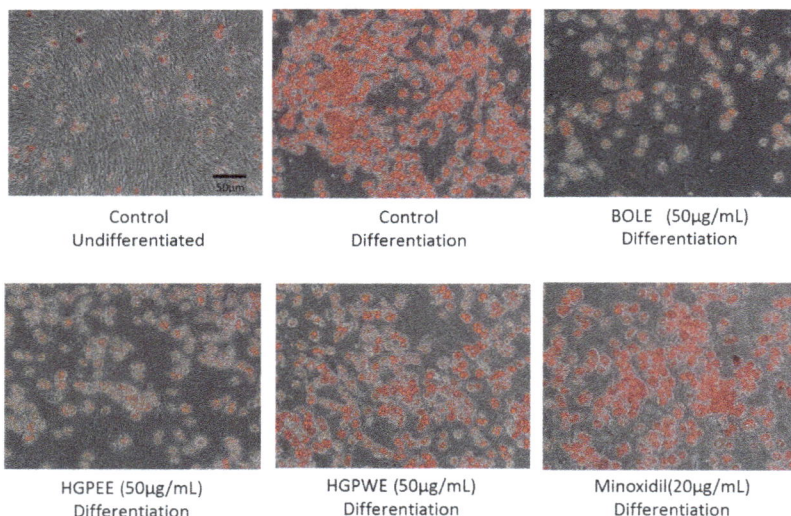

Control
Undifferentiated

Control
Differentiation

BOLE (50μg/mL)
Differentiation

HGPEE (50μg/mL)
Differentiation

HGPWE (50μg/mL)
Differentiation

Minoxidil(20μg/mL)
Differentiation

Figure 2. The shape of microscopic features of sebocytes dyed in oil red O.

3.2. Effects of BOLE, HGPEE, HGPWE, and Minoxidil on the Growth of Human Fetal Epidermal Keratinocytes in a Two-Layer Culture System of Human Hair Papilla Cells and Human Epidermal Keratinocytes

Figure 3 shows the results of the effects of BOLE, HGPEE, HGPWE, and minoxidil on the growth of human epidermal keratinocytes in a two-layer culture system of human hair papilla cells and human fetal epidermal keratinocytes. After 5 days of culture, the number of cells significantly increased after treatment with HGPEE at 12.5, 25, and 50 μg/mL relative to that after treatment with control. Minoxidil also significantly increased the number of human fetal epidermal keratinocytes relative to that of control at 10 μg/mL, but not at 5 or 20 μg/mL. No significant difference was observed in the number of human fetal epidermal keratinocytes between BOLE and the control at 12.5, 25, and 50 μg/mL. Additionally, no significant difference in the number of cells was seen after treatment with HGPWE at 12.5, 25, and 50 μg/mL.

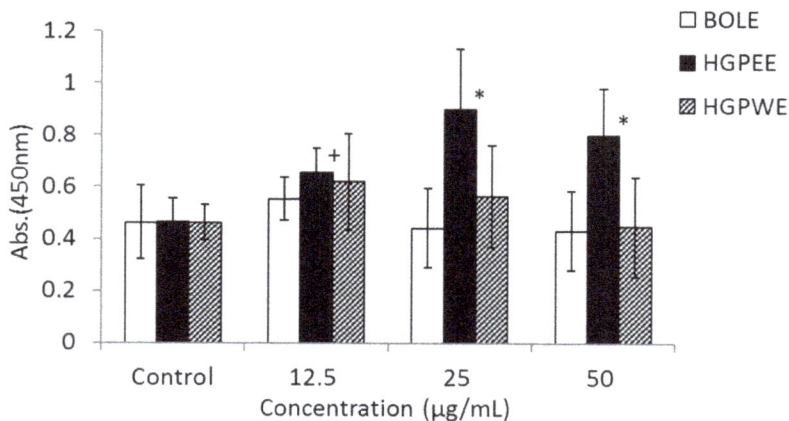

Figure 3. *Cont.*

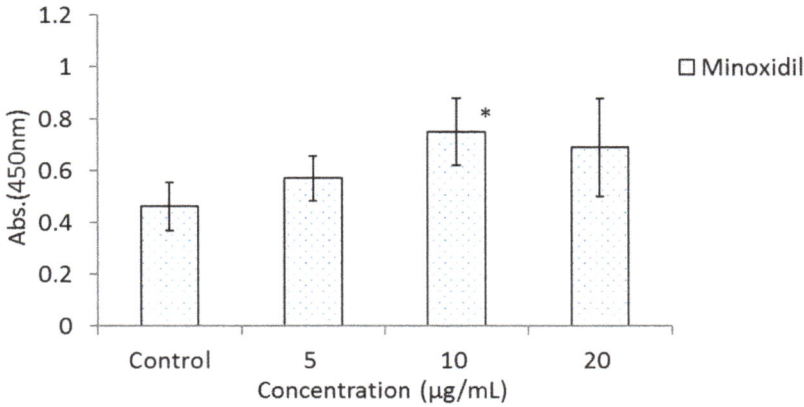

Figure 3. Effects of BOLE, HGPEE, HGPWE and minoxidil on the number of human fetal epidermal keratinocytes in a two-layer culture of human hair papilla cells and human epidermal keratinocytes. n = 4, mean ± S.D. * $p < 0.05$ vs. Control, $^{+}$ $p < 0.1$ vs. Control.

3.3. Effects of HGPEE, HGPWE, and Minoxidil on Expression of mRNA in Human Fetal Epidermal Keratinocytes Related to Hair Growth

The results of investigating the effect of each sample on the mRNA expressions of Wnt10b, Shh, and Lef1 among hair growth-related factors in the cultured human epidermal keratinocytes are shown in Figure 4. The mRNA expression of Wnt10b increased 2.6-fold, 1.8-fold, and 1.6-fold by HGPEE (50 μg/mL), HGPWE (50 μg/mL), and minoxidil (20 μg/mL), respectively, and the differences were significant between HGPEE (50 μg/mL) or HGPWE (50 μg/mL) and the control. A tendency to increase Wnt10b mRNA expression was observed upon treatment with minoxidil. Wnt10b is expressed by epidermal keratinocytes, and it induces translocation of β-catenin to the nucleus to form a complex with Lef-1 that induces transcription of downstream target genes. Lef1 is a transcription factor that works downstream in the Wnt signaling pathway, and participates in hair follicle regeneration. Lef1 increased 2.3 times by HGPEE (50 μg/mL), which was significantly different from that of the control. No significant difference in Lef1 was seen by HGPWE (50 μg/mL) or minoxidil (20 μg/mL) and the control. Moreover, although expression of Shh increased 1.6 times by HGPEE (50 μg/mL), a significant difference was not observed as the standard deviation was very high. No significant difference in Shh between HGPWE (50 μg/mL) or minoxidil (20 μg/mL) and the control was observed.

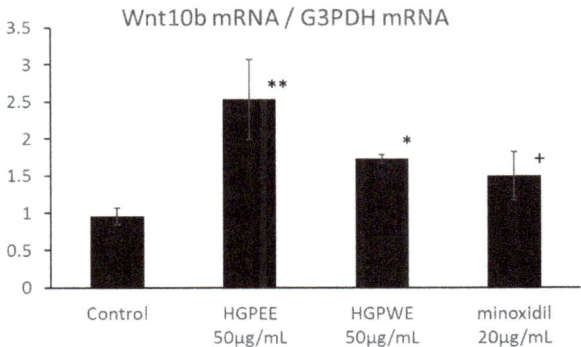

Figure 4. *Cont.*

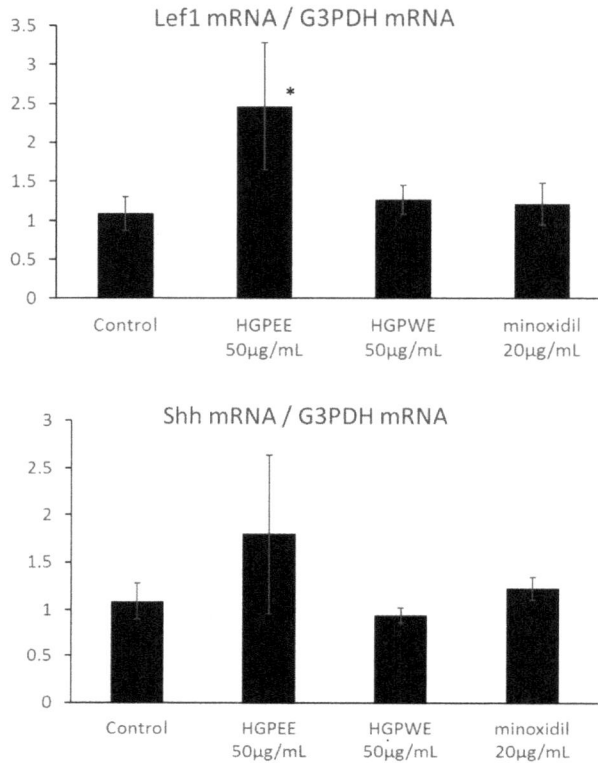

Figure 4. Effect of HGPEE, HGPWE, and minoxidil on the mRNA expressions of Wnt10b, Lef1, and Shh in cultured human fetal epidermal keratinocytes. n = 3, mean ± S.D., * $p < 0.05$ vs. Control, ** $p < 0.01$ vs. Control, + $p < 0.1$ vs. Control.

3.4. Effects of HGPEE, HGPWE, and Minoxidil on the mRNA Expressions of Hair Growth-Related Factors in Human Hair Papilla Cells

Hair papilla cells play an important role in hair follicle formation. Therefore, we investigated the effects of HGPEE, HGPWE, and minoxidil on the mRNA expressions of hair growth-related factors such as KGF and VEGF in human hair papilla cells. Figure 5 shows the results of investigating the effect of each sample on the mRNA expressions of KGF and VEGF among hair growth-related factors in cultured human papilla cells. The amount of mRNA of KGF was increased by HGPEE (12.5, 25 µg/mL) in a concentration-dependent manner, and a significant difference in the amounts was observed between HGPEE (12.5 µg/mL) and the control. The amount of KGF mRNA was not affected by HGPWE (50 µg/mL) or minoxidil (20 µg/mL). The amount of VEGF mRNA tended to increase in a concentration-dependent manner with treatment by HGPEE (12.5, 25 µg/mL). The amount of VEGF mRNA was not influenced by HGPWE (50 µg/mL) or minoxidil (20 µg/mL).

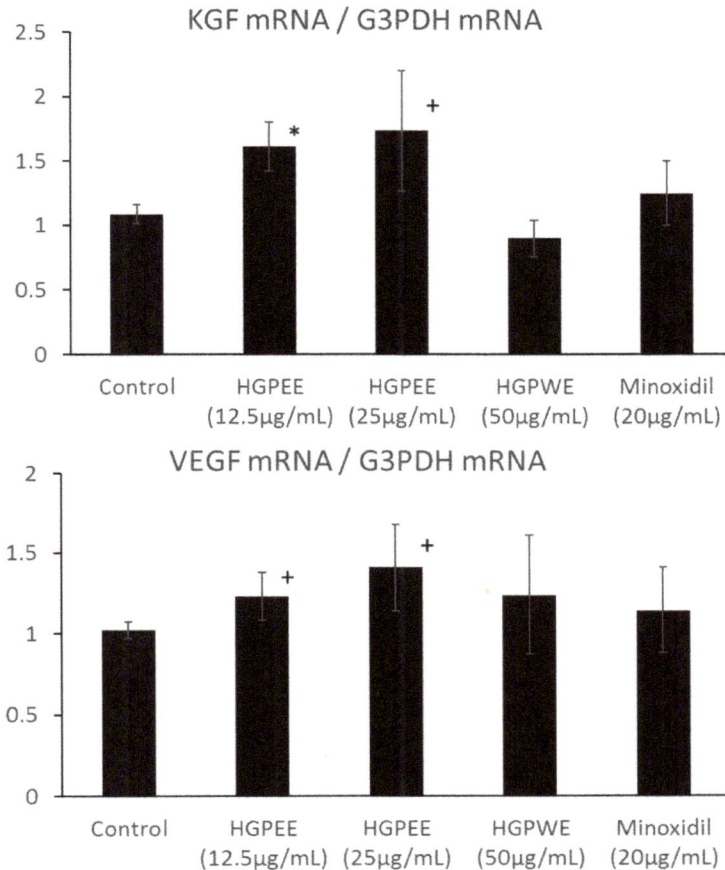

Figure 5. Effect of HGPEE, HGPWE, and minoxidil on the mRNA expressions of KGF, and VEGF in cultured human hair papilla cells. n = 3, mean ± S.D., * $p < 0.05$ vs. Control, + $p < 0.1$ vs. Control.

4. Discussion

The scalp has more sebaceous glands than other parts of the skin, and these glands exhibit a high amount of holocrine secretion of sebum. When a sebaceous gland becomes enlarged and sebum high, hair loss can result in seborrheic alopecia. To close pores, scalp sebum is secreted from sebaceous glands more than is necessary, and inflammation occurs around the hair roots. This can lead to hair loss and seborrheic alopecia. Excessive sebum secretion and indigestion are known causes of hair loss. As BOLE and HGPEE decreased the amount of sebum per cell relative to that by the control, it is thought that they could be useful for preventing hair loss caused by excessive scalp sebum secretion in males and females. HGPEE, especially, strongly decreased the size of lipid droplets, it is necessary to identify the active ingredient in BOLE and HGPEE that prevents depilation caused by an excess of scalp sebum. Sebum mainly consists of a triglyceride (fatty acid ester of glycerin) with lesser amounts of squalene and waxes. When sebum clogs skin pores, fat caps form over the pores, and a true fungus (*Malassezia*) multiplies by using triglyceride as nutrition and causes inflammation. BOLE can prevent adverse effects caused by excessive sebum in the scalp as it has been reported to have an antibacterial effect [17–20] and an anti-inflammatory effect of 15-methoxypinusolidic acid is contained in BOLE [21].

Dihydrotestosterone, a male hormone, stimulates a high production of sebum in the sebaceous gland and is secreted in hair pores. As male pattern baldness is a disease caused by excessive sensitivity to dihydrotestosterone, it causes symptoms in young and middle-aged persons similar to sebum-related symptoms of the scalp. In baldness that develops in young people, the period of anagen (hair growth phase) is shortened by male hormones so only thin and very short hair increases instead of thick and long hair. Excessive sebum is considered to be one of the causes of premature balding. The seed of *T. occidentalis*, which is closely related to *B. orientalis* L., causes 5α-reductase inhibition [16] via actions of the active ingredients, flavonoid and diterpene, which have been identified in BOLE [22–24]. BOLE causes 5α-reductase inhibition and is expected to cause strong inhibition of dihydrotestosterone from testosterone in hair papilla cells and prevent excessive sebum.

On the other hand, hair grows thickly for a long time as the hair follicle cells in the hair root repeat division and push up cells one after another. Hair growth undergoes a cycle (anagen phase→catagen phase→telogen phase), and it is thought that a hair papilla cell present at the root of the hair controls this cycle.

The mechanism of hair growth has not been elucidated in detail, but it is thought that activation of hair papilla cells or hair follicle cells is important. However, if Wnt10b is activated, smaller hair follicles increase in size, and if KGF is promoted, the growth phase of hair will be extended. These two effects should promote hair growth/restoration.

When the organ culture of mice mustache hair follicles was performed, Wnt10b promoted hair shaft growth [7]. Although other Wnt families (Wnt3a, Wnt5a, Wnt11) (except for Wnt10b) have not caused significant extension, Wnt10b is thought to promote the cell growth of hair follicles [7]. Moreover, immunohistochemical analysis of frozen sections of hair follicles has shown that proliferation of lower hair follicle cells was indirectly promoted remarkably by Wnt10b and that this effect involved the β-catenin pathway [25]. Specifically, it was suggested that Wnt10b functions as a specific growth and differentiation-enhancing factor in hair follicles. Thus, the Wnt/β-catenin pathway is closely involved in the formation of hair follicles. The Shh molecule signal has an important role in hair follicle formation; however, how the process is controlled remains unknown. A study has shown that hair did not grow following incomplete development of papilla cells in the skin of Shh mutant mice, which indicated that Shh may control the proliferation of hair follicles [6]. Although investigation of epimorphin, an agonist of Shh, as a hair growth agent has been shown to induce hair follicle formation and promote the growth phase, development was discontinued due to safety concerns [26]. Furthermore, it has been reported that continuous β-catenin signaling may cause hair follicle tumors [27]. HGPEE is a plant mixture ethanol extract that has been used for years, and because it does not activate Shh directly, it is thought that there are no safety problems.

In addition, it is thought that minoxidil suppresses apoptosis of hair follicle cells by improving activation of SUR (sulfonylurea receptor) and mitochondrial ATP-dependent K-channel opening, and also increases the promotion of hair papilla cells and hair organization blood-flow improvement action caused by vascular smooth muscle ATP sensitivity K-channel opening [28]. In mice where the epithelium system strongly expressed VEGF in the presence of a K14 promoter, increases in blood vessels around hair follicles have been observed, and longer than usual thick hair has been recognized [29]. From these observations, it is thought that VEGF produced from hair papilla cells has a hair growth effect via increased action on the blood vessels near the hair follicles, which improves hair organization blood-flow. As the amount of mRNA of Wnt10b increased by HGPEE, as were the amounts of mRNA of KGF and VEGF, prevention of lipid accumulation in hair pores and hair growth promotion and hair restoration effects of hair restorer products containing HGPEE are expected. Furthermore, it has been reported that excessive activation of N-methyl-D-aspartate (NMDA) causes apoptosis of hair cells [30]. 15-Methoxypinusolidic acid (15-MPA) contained in HGPEE has been reported to inhibit glutamate-induced increase in intracellular calcium ($[Ca^{2+}]_i$), which prevented NMDA-induced cytotoxicity [31]. Moreover, 15-MPA has been shown to successfully

Cosmetics **2017**, *4*, 29

reduce subsequent overproduction of nitric oxide, reduce cellular peroxide concentrations, and inhibit glutathione depletion and lipid peroxidation induced by glutamate in cultures [31].

Based on previous results, it was possible to conclude that the ethanol extract (HGPEE) was more effective than the aqueous one (HGPWE) in the extraction of hair growth-related factors. Both the leaves and seeds contained an essential oil consisting of borneol, bornyl acetate, thujone, camphor, and sesquiterpenes [32]. The leaves also contained rhodoxanthin, amentoflavone, quercetin, myricetin, carotene, xanthophyll, and ascorbic acid [32]. Many fatty acids and nonpolar substances were present in the ethanol plant mixture extract. Further research and clinical tests are required to identify which ingredients in HGPEE contributed to the hair growth/restoration functionality demonstrated in this study.

5. Conclusions

In conclusion, the results from this study showed that BOLE reduced generation of sebum from sebocytes and decreased the size of lipid droplets. Moreover, an ethanol extract containing *B. orientalis* and other plant materials (HGPEE) increased the amounts of mRNA of Wnt10b and Lef1 in cultured human fetal epidermal keratinocytes, as well as the amounts of mRNA of VEGF and KGF in cultured human hair papilla cells. These results indicate that future studies should focus on the timing of hair cycle phases, specifically to learn if HPGEE lengthens the anagen phase in specific hair follicle cells to promote hair growth.

Acknowledgments: The author gratefully acknowledges the technical assistance of Rin Kouriki, Yuuki Matsuzaki and Kyouhei Igarashi.

Author Contributions: H.Z., L.G. and K.M. performed the experiments. K.M. designed the study and performed the analysis. H.Z., L.G., and K.M. interpreted the data and drafted the manuscript. K.M. supervised the progress and critically revised the manuscript. All authors read and approved the final manuscript.

Conflicts of Interest: The authors declare that they have no competing interests.

Ethics Approval and Consent to Participate: Not applicable.

Funding: This study was supported by a grant from Asian Scalp Healthy Research Center in Kobe, Japan.

References

1. Kellum, R.E. Human sebaceous gland lipids. Analysis by thin-layer chromatography. *Arch. Dermatol.* **1967**, *95*, 218–220. [CrossRef] [PubMed]
2. Ito, A.; Sakiguchi, T.; Kitamura, K.; Akamatsu, H.; Horio, T. Establishment of a tissue culture system for hamster sebaceous gland cells. *Dermatology* **1988**, *197*, 238–244. [CrossRef]
3. Sato, T.; Imai, N.; Akimoto, N.; Sakiguchi, T.; Kitamura, K.; Ito, A. Epidermal growth factor and 1alpha,25-dihydroxyvitamin D3 suppress lipogenesis in hamster sebaceous gland cells in vitro. *J. Investig. Dermatol.* **2001**, *117*, 965–970. [CrossRef] [PubMed]
4. Millar, S.E. Molecular mechanisms regulating hair follicle development. *J. Investig. Dermatol.* **2002**, *118*, 216–225. [CrossRef] [PubMed]
5. Reddy, S.; Andl, T.; Bagasra, A.; Lu, M.M.; Epstein, D.J.; Morrisey, E.E.; Millar, S.E. Characterization of Wnt gene expression in developing and postnatal hair follicles and identification of Wnt5a as a target of Sonic hedgehog in hair follicle morphogenesis. *Mech. Dev.* **2001**, *107*, 69–82. [CrossRef]
6. Chiang, C.; Swan, R.Z.; Grachtchouk, M.; Bolinger, M.; Litingtung, Y.; Robertson, E.K.; Cooper, M.K.; Gaffield, W.; Westphal, H.; Beachy, P.A.; et al. Essential role for Sonic hedgehog during hair follicle morphogenesis. *Dev. Biol.* **1999**, *205*, 1–9. [CrossRef] [PubMed]
7. Ouji, Y.; Yoshikawa, M.; Moriya, K.; Nishiofuku, M.; Matsuda, R.; Ishizaka, S. Wnt-10b, uniquely among Wnts, promotes epithelial differentiation and shaft growth. *Biochem. Biophys. Res. Commun.* **2008**, *367*, 299–304. [CrossRef] [PubMed]
8. Van Genderen, C.; Okamura, R.M.; Fariñas, I.; Quo, R.G.; Parslow, T.G.; Bruhn, L.; Grosschedl, R. Development of several organs that require inductive epithelial-mesenchymal interactions is impaired in LEF-1-deficient mice. *Genes Dev.* **1994**, *8*, 2691–2703. [CrossRef] [PubMed]

9. Goldman, C.K.; Tsai, J.C.; Soroceanu, L.; Gillespie, G.Y. Loss of vascular endothelial growth factor in human alopecia hair follicles. *J. Investig. Dermatol.* **1995**, *104*, 18S–20S. [CrossRef] [PubMed]
10. Lachgar, S.; Moukadiri, H.; Jonca, F.; Charveron, M.; Bouhaddioui, N.; Gall, Y.; Bonafe, J.L.; Plouët, J. Vascular endothelial growth factor is an autocrine growth factor for hair dermal papilla cells. *J. Investig. Dermatol.* **1996**, *106*, 17–23. [CrossRef] [PubMed]
11. Lachgar, S.; Charveron, M.; Ceruti, I.; Lagarde, J.M.; Gall, Y.; Bonafe, J.L. VEGF mRNA expression in different stages of the human hair cycle, analysis by confocal laser microscopy. In *Hair Research for the Next Millennium*; Van Neste, D., Randall, V.A., Eds.; Elsevier Science: Amsterdam, The Netherlands, 1996; p. 407.
12. Guo, L.; Degenstein, L.; Fuchs, E. Keratinocyte growth factor is required for hair development but not for wound healing. *Genes Dev.* **1996**, *10*, 165–172. [CrossRef] [PubMed]
13. Danilenko, D.M.; Ring, B.D.; Yanagihara, D.; Benson, W.; Wiemann, B.; Starnes, C.O.; Pierce, G.F. Keratinocyte growth factor is an important endogenous mediator of hair follicle growth, development, and differentiation. Normalization of the nu/nu follicular differentiation defect and amelioration of chemotherapy-induced alopecia. *Am. J. Pathol.* **1995**, *147*, 145–154. [PubMed]
14. Yeung, H.C. *Handbook of Chinese Herbs and Formulas*; Institute of Chinese Medicine: Los Angeles, CA, USA, 1985.
15. Zhang, N.N.; Park, D.K.; Park, H.J. Hair growth-promoting activity of hot water extract of *Thuja orientalis*. *BMC Complement Altern. Med.* **2013**, *13*, 9. [CrossRef] [PubMed]
16. Park, W.S.; Lee, C.H.; Lee, B.G.; Chang, I.S. The extract of *Thujae occidentalis* semen inhibited 5alpha-reductase and androchronogenetic alopecia of B6CBAF1/j hybrid mouse. *J. Dermatol. Sci.* **2003**, *31*, 91–98. [CrossRef]
17. Jain, R.K.; Garg, S.C. Antimicrobial activity of the essential oil of *Thuja orientalis* L. *Ancient. Sci. Life* **1997**, *16*, 186–189.
18. Hassanzadeh, M.K.; Rahimizadeh, M.; Fazly Bazzaz, B.S.; Emami, S.A.; Asili, J. Chemical and antimicrobial studies of Platycladus orientalis essential oils. *Pharm. Biol.* **2001**, *5*, 388–390. [CrossRef]
19. Duhan, J.S.; Saharan, P.; Surekha; Kumar, A. Antimicrobial potential of various fractions of *Thuja orientalis*. *Afr. J. Microbiol. Res.* **2013**, *7*, 3179–3186.
20. Jasuja, N.D.; Sharma, S.K.; Saxena, R.; Choudhary, J.; Sharma, R.; Joshi, S.C. Antibacterial, antioxidant and phytochemical investigation of *Thuja orientalis* leaves. *J. Med. Plants Res.* **2013**, *7*, 1886–1893.
21. Choi, Y.; Moon, A.; Kim, Y.C. A pinusolide derivative, 15-methoxypinusolidic acid from Biota orientalis inhibits inducible nitric oxide synthase in microglial cells, implication for a potential anti-inflammatory effect. *Int. Immunopharmacol.* **2008**, *8*, 548–555. [CrossRef] [PubMed]
22. Zhu, J.X.; Wang, Y.; Kong, L.D.; Yang, C.; Zhang, X. Effects of Biota orientalis extract and its flavonoid constituents, quercetin and rutin on serum uric acid levels in oxonate-induced mice and xanthine dehydrogenase and xanthine oxidase activities in mouse liver. *J. Ethnopharmacol.* **2004**, *93*, 133–140. [CrossRef] [PubMed]
23. Lu, Y.; Liu, Z.; Wang, Z.; Wei, D. Quality evaluation of *Platycladus orientalis* (L.) Franco through simultaneous determination of four bioactive flavonoids by high-performance liquid chromatography. *J. Pharm. Biomed. Anal.* **2006**, *4*, 1186–1190. [CrossRef] [PubMed]
24. Duhan, J.S.; Saharan, P.; Gahlawat, S.K. Antioxidant potential of various extracts of stem of Thuja orientalis, in vitro study. *Intern. J. App. Bio Pharma. Technol.* **2013**, *3*, 264–271.
25. Zhang, Y.; Xing, Y.; Guo, H.; Ma, X.; Li, Y. Immunohistochemical study of hair follicle stem cells in regenerated hair follicles induced by Wnt10b. *Int. J. Med. Sci.* **2016**, *13*, 765–771. [CrossRef] [PubMed]
26. Takebe, K.; Oka, Y.; Radisky, D.; Tsuda, H.; Tochigui, K.; Koshida, S.; Kogo, K.; Hirai, Y. Epimorphin acts to induce hair follicle anagen in C57BL/6 mice. *FASEB J.* **2003**, *17*, 2037–2047. [CrossRef] [PubMed]
27. Lo Celso, C.; Prowse, D.M.; Watt, F.M. Transient activation of beta-catenin signalling in adult mouse epidermis is sufficient to induce new hair follicles but continuous activation is required to maintain hair follicle tumours. *Development* **2004**, *131*, 1787–1799. [CrossRef] [PubMed]
28. Li, M.; Marubayashi, A.; Nakaya, Y.; Fukui, K.; Arase, S. Dermal papilla cells possess sulfonylurea receptor 2B and adenosine receptors which are the possible mediators of minoxidil-induced VEGF productions. *J. Investig. Dermatol.* **2001**, *117*, 1594–1600. [PubMed]
29. Yano, K.; Brown, L.F.; Detmar, M. Control of hair growth and follicle size by VEGF-mediated angiogenesis. *J. Clin. Investig.* **2001**, *107*, 409–417. [CrossRef] [PubMed]

30. Sheets, L. Excessive activation of ionotropic glutamate receptors induces apoptotic hair-cell death independent of afferent and efferent innervation. *Sci. Rep.* **2017**, *7*, 41102. [CrossRef] [PubMed]
31. Koo, K.A.; Kim, S.H.; Lee, M.K.; Kim, Y.C. 15-Methoxypinusolidic acid from Biota orientalis attenuates glutamate-induced neurotoxicity in primary cultured rat cortical cells. *Toxicol. In Vitro* **2006**, *20*, 936–941. [CrossRef] [PubMed]
32. Bown, D. *Encyclopaedia of Herbs and Their Uses*; Dorling Kindersley: London, UK, 1995.

cosmetics

MDPI

Article

In Vitro and In Vivo Evaluation of Nanoemulsion Containing Vegetable Extracts

Pedro Alves Rocha-Filho [1,*], Marcio Ferrari [2], Monica Maruno [3], Odila Souza [1] and Viviane Gumiero [1]

[1] Department of Pharmaceutical Sciences, Faculty of Pharmaceutical Sciences of Ribeirão Preto, University of São Paulo, Avenida do Café, s/n, Bairro Monte Alegre, Ribeirão Preto, SP 14040-903, Brazil; odilinha@yahoo.com.br (O.S.); vcgumiero@hotmail.com (V.G.)

[2] College of Pharmacy, Federal University of Rio Grande do Norte, Rua Gustavo Cordeiro de Farias, s/n, Petrópolis, Natal, RN 59012-570, Brazil; ferrarimarcio@uol.com.br

[3] Pharmacy Course Coordination, Centro Universitário Barão de Mauá, R. Ramos de Azevedo, 423, Jardim Paulista, Ribeirão Preto, SP 14090-180, Brazil; monica.maruno@baraodemaua.br

* Correspondence: pedranjo@fcfrp.usp.br; Tel.: +55-16-3315-4214

Received: 19 July 2017; Accepted: 29 August 2017; Published: 7 September 2017

Abstract: Oil/Water nanoemulsions were obtained, employing PEG castor oil derivatives/fatty esters surfactant, babassu oil, and purified water from a study based on phase diagrams. The nanoemulsions had been prepared by a low energy process inversion phase emulsion. Different parameters, such as order of addition of the components, temperature, stirring speed, and time, were studied to prepare O/W nanoemulsions. The influence of vegetable extract addition on size distribution of nanoemulsions was also analyzed. Evaluation of the nanoemulsions was studied in vitro by HET-CAM and RDB methods. Stable transparent bluish O/W babassu oil nanoemulsion were obtained with surfactant pair fatty ester/PEG-54 castor oil, in an $HLB_{required}$ value = 10.0 and with a particle droplet size of 46 ± 13 nm. Vegetable extract addition had not influenced nanoemulsion's stability. The results obtained for in vitro and in vivo nanoemulsion evaluation, based on the hydration and oiliness, and pH of the skin, shows O/W nanoemulsions as potential vehicle for topical application.

Keywords: nanoemulsions; babassu oil; vegetable extract; efficacy evaluation

---

1. Introduction

For cosmetics products, nanoemulsions are preferable and more stable than macroemulsions, have good spreadability, and facilitate penetration of actives into the skin. The interest in nanoscale emulsion has been growing considerably in recent decades, due to its specific attributes, such as high stability, attractive appearance and drug delivery properties; therefore, performance is expected to improve using a lipid-based nanocarrier.

Nanoemulsions have recently become increasingly important as potential vehicles for the controlled delivery of cosmetics, and for the optimized dispersion of active ingredients, in particular, skin layers. Due to their lipophilic interior, nanoemulsions are more suitable for the transport of lipophilic compounds than liposomes. Similar to liposomes, they support the skin penetration of active ingredients, and thus, increase their concentration in the skin. Another advantage is the small-sized droplet, with its high surface area, allowing effective transport of the active to the skin. Furthermore, nanoemulsions gain increasing interest due to their own bioactive effects. This may reduce the transepidermal water loss (TEWL), indicating that the barrier function of the skin is strengthened. Nanoemulsions are acceptable in cosmetics, because there is no inherent creaming, sedimentation, flocculation, or coalescence that is observed with macroemulsions. Nanoemulsions are generated by different approaches: the so-called high-energy and low-energy methods.

Pereira et al. [1] used the oils of *Rubus idaeus* (raspberry seed oil), *Passiflora edulis* (seed oil), and *Prunus persica* (peach kernel oil), associated with lanolin derivatives, for Oil/Water nanoemulsions obtention. The authors verified that these lanolin derivatives cause alterations in the particle size, but they can be used for topical application.

Most recently Rocha-Filho et al. [2] had studied the action of tea tree oil and lavender oil on the stability of rice bran oil nanoemulsions. The authors observed that the presence of lavender essential oil causing a decrease in particle size, so, promoting the stabilization of the dispersed system.

The purpose of this article was to produce nanoemulsions, containing babassu oil and different vegetables extracts, through a low-energy method. So, babassu oil (*Orbignya oleifera*) was the oil phase, and is used in the treatment of various skin disorders, due to its anti-inflammatory, antiseptic, and healing properties. The aqueous phase was composed for *Areca catechu* seed extract, *Glycyrrhiza glabra* root extract (licorice), *Portulaca oleracea* (portulaca) extract and purified water. These extracts were recommended for skin recovery, mainly because of their antioxidant, anti-inflammatory, antibacterial, astringent, anti-hyaluronidase, melanogenesis inhibitor, anti-irritant, and healing promoter activities.

This study had several objectives: (1) to develop nanoemulsions containing babassu oil; (2) to incorporate vegetable extracts in the formulations, such as *Areca catechu* seed extract (Ar), *Glycyrrhiza glabra* (Al) and *Portulaca oleracea* extract; (3) to perform in vitro nanoemulsion evaluation by HET-CAM and RDC methods; and (4) to evaluate nanoemulsions' effect on skin hydration, oiliness, and pH, in vivo.

2. Materials

Oil phase: *Orbignya oleifera* seed oil (Crodamazon® Babaçu CO-Croda Brasil); surfactants: sorbitan monooleate (Span® 80) (HLB value = 4.3); PEG-54 castor oil (Ultramona® R540) (Hydrophilic Lipophilic Balance value = 14.4) (all surfactants Oxiteno Brasil); vegetable hydrophilic extracts: *Areca catechu* seed extract; *Glycyrrhiza glabra* (licorice) root extract and *Portulaca oleracea* extract (employed as provided by Lipo do Brasil); and aqueous phase: purified water.

3. Methods

3.1. Formulation Studies

3.1.1. Nanoemulsion Development

The emulsions were prepared by the emulsion phase inversion method (EPI). The aqueous and oil phases were heated separately at 75 ± 2 °C. Then, the aqueous phase was added slowly over to the oily phase containing surfactants, and constant stirring (600 rpm) (Mod-Fisaton mechanical stirrer-713, Fisaton, São Paulo, SP, Brasil) until mixture reaches room temperature (25 ± 2 °C) [3].

3.1.2. Vegetable Extract Addition on Babassu Oil Nanoemulsions

Nanoemulsions were prepared by the cited method (3.1.1) and vegetable extracts were added separately at the recommended concentration used in scientific literature, 3.0%.

3.1.3. Droplet Size and Polydispersity Index Determination

Droplet size diameter and polydispersity index of the nanoemulsions were determined by dynamic light scattering (DLS) (Nanosizer Malvern ZS, Worcestershire, UK) at a scattering angle of 173°, and the samples were diluted in purified water in the proportion 1:100, at 25 °C.

3.1.4. pH Evaluation

Nanoemulsion (1.0 g) was homogenized with 9.0 g of purified water, and the pH value was measured by inserting the electrode (pH meter Analion-Mod. PM608, Analion, Ribeirão Preto, SP, Brazil) directly into the sample solution at 24.0 ± 2.0 °C [4].

3.1.5. Electrical Conductivity Evaluation

The electrical conductivity of the emulsions was evaluated with the conductivity Digimed (DM32, Digimed, São Paulo, Brazil) calibrated with standard solution: the electrode was inserted directly on the sample at 24.0 ± 2.0 °C [4,5].

3.1.6. Refractive Index

The RI of the system was measured (in triplicates) by an Abbe refractometer (Bausch and Lomb Optical Company, Rochester, NY, USA) by placing one drop of the formulation on the slide, at 24 ± 2 °C. The refractometer was calibrated with purified water (1.333) [6].

3.1.7. Freeze-Defrost Cycles

The nanoemulsions were subjected to a temperature of 45 ± 5 °C for 24 h, and after, to a temperature of 4 ± 2 °C also for 24 h, thereby completing a cycle. Macroscopic evaluation was made 24 h after nanoemulsions' preparation, and at the end of the 6th cycle (day 12) [7].

3.2. In Vitro Nanoemulsion Evaluation

3.2.1. Hen's Egg Test on the Chorioallantoic Membrane (HET-CAM)

This method corresponds to a modification of the method described by [8]: embryonated chicken eggs were purchased from commercial hatcheries 10 days after fertilization and were maintained at 37 ± 2 °C. With scissors, a small incision was made in the center of the upper part of the shell of the egg. Subsequently, the whitish membrane that adhered to the inside of the egg was pulled out, to make clear the chorioallantoic membrane transparent with blood vessels.

The samples (0.3 g) were applied on the egg chorioallantoic membrane, and after 20 s of contact, it was washed with saline solution (5 mL). It was determined, the time (seconds) at which signs of irritation appear. Sodium lauryl sulfate (SLS) (10.0% solution) was used as a positive control, and saline solution (0.9%) and purified water, both, as a negative control.

The irritations produced were evaluated according to Luepke scale (Table 1), and were classified according Table 2. The test for each sample was performed in quadruplicate.

Table 1. Luepke scale for the appearance of the phenomena as function of time.

Phenomena	Time (T)		
	T ≤ 30 s	30 s > T ≤ 2 min	2 min > T ≤ 5 min
Hyperemia	5	3	1
Hemorrhage	7	5	3
Coagulation	9	7	5

Table 2. Classification of products according to the scores of the phenomena.

HET-CAM Index	Classification
N ≤ 1	practically not irritant
1 < N ≥ 5	slightly irritant
5 < N ≥ 9	moderately irritant
N > 9	irritant

3.2.2. Irritation Test in Red Blood Cell (RBC) System Cellular Model

The human venous blood samples were freshly collected and put into a test tube containing anticoagulant (EDTA-Na$_2$ 10.0%) (CEP/FCFRP n^r 204/2011). For the calculation of H$_{50}$ (effective concentration that causes 50.0% of hemolysis), formulations and surfactant solution separately were diluted (triplicates) to 0.02, 0.05, 0.1, 0.2, 0.4, 0.5, 0.6, 0.7, and 0.8 g/mL, in phosphate buffered saline at 10.0%, then, added 50.0 µL of blood, and the tubes were homogenized and incubated for 90 min at room temperature [9]. The samples were centrifuged at 2000 rpm for 5 min, and the supernatant was removed to measure the absorbance (540 nm) against the blank (100.0% buffered solution containing red blood cells). The results were compared with a tube in which the cells were completely lysed by distilled water (positive control). The hemolytic activity of each sample was calculated by the formula:

$$H(\%) = \frac{Abs_{sample}}{Abs_{control}}^+$$

For hemolytic activity, samples with values higher than H$_{50}$ were analyzed based on red blood cells denaturation. The absorbance readings of the supernatants are performed at wavelengths 540 and 575 nm against the blank (test substance diluted in buffer). The value of the extinction measured at 575 nm (α) is divided by the value of extinction measured at 540 nm (β) to obtain the ratio α/β. This ratio is used to characterize the denaturing index (DI) of hemoglobin.

$$DI(\%) = (R_1 - R_i)/(R_1 - R_2) \times 100$$

where:

R$_1$—ratio α/β hemoglobin;
R$_i$—ratio α/β of the test substance;
R$_2$—ratio α/β 0.1% of the SLS.

The relationship between the concentration that causes 50.0% hemolysis (H$_{50}$) and the denaturing index (DI) is defined as the ratio of H$_{50}$/ID, and was calculated for each test substance; the irritation potential is classified according to Table 3.

Table 3. Red blood cell (RBC) system cellular model applied to determine potential irritation of nanoemulsions [9].

Ratio (H$_{50}$/ID)	Classification
>100.0	No irritant
≥10.0	Slightly irritant
≥1.0	Moderately irritant
≥0.1	Severe irritant
<0.1	Maximum irritant

3.3. In Vivo Nanoemulsion Evaluation

3.3.1. Anti-Inflammatory Activity Evaluation by Edema Ear Rats Induced by Croton Oil

The mice used were kind of "Swiss" females, weighing between 25–30 g. The animals were kept in light/dark cycles of 12 h, with access to water and feed during the experiment. The number of animals used in the experiment was 38, divided into 5 groups of 6 to 8 animals each. The project was approved by the Ethics Committee on Animals (Protocol n^r 10.1.1180.53.5-FCFRP-USP, 02/04/2011).

The anti-inflammatory activity was tested by acute ear edema, after topical application of 20 µL of croton oil solution (5.0% v/v) in acetone on the inner surface of the right ear (the left ear was correspondent control to right ear), with the exception of group I (negative control), which received no treatment. Thirty minutes after croton oil application (20 μL), groups II, III, IV and V were treated with only the vehicle, babassu oil nanoemulsions without vegetable extracts, babassu oil nanoemulsions with vegetable extracts, and dexamethasone solution (positive control—4.0 mg/mL), respectively. In groups I, II, and V, acetone (20 μL) was applied in the left ear, while acetone (20 μL) was applied for groups III and IV over formulation's vehicle (20 μL) (this vehicle was composed of water, surfactants, microbial preservative, and BHT).

Four hours later, the mice were sacrificed and the ear thickness was measured (mm). Then, ear circles of 6 mm were cut out and weighed on an analytical balance, to assess the intensity of mass edema (mg). In both methods, thickness and weight values for the right ears were discounted from the opposite side ears in all groups, and converted to a percentage, relative to the negative control [10,11].

3.3.2. Skin Hydration, Oiliness, and pH Evaluation

After approval by the Ethics Committee on Human (Protocol n^r 204-FCFRP-USP, 04/02/2011, 07/09/2011), babassu oil nanoemulsion dermal activity assessment was held in a room with controlled relative humidity ($60 \pm 3\%$) and temperature ($22 \pm 2\,^{\circ}$C).

(a) *Inclusion and exclusion criteria for selecting volunteers* [12].

It was selected 30 healthy volunteers, female and male, aged between 18 and 30 years, and free of skin care product use for at least 30 days. Volunteers with skin diseases or hypersensitivity to any component of the formulation were not accepted.

(b) *Application of formulations*

Six areas were defined on each volunteer's forearm, totaling 12 areas. The demarcated areas were divided according to the following groups (triplicate):

(1) babassu oil nanoemulsion without plant extracts;
(2) babassu oil nanoemulsion with plant extracts;
(3) commercial nanoemulsion Mist Moisturizing Milk Ekos®Cupuaçu (Natura, São Paulo, Brazil);
(4) control areas (without application of any formulation).

Babassu oil nanoemulsion (50.0 mg) was applied in a circular motion in a specific site of volunteer's forearm skin, and the assessment was made after 30, 60, 90, and 120 min. Skin hydration, pH, and oiliness measurements were performed in triplicate in each demarcated area with Corneometer® CM 820 equipment (Courage + Khazaka Electronic GmbH, Cologne, Germany) [13].

3.3.3. Skin Hydration Evaluation

The changes in capacitance were detected by a probe, and converted to hydration units ranging 0–150 arbitrary units (AU), where 0 corresponds to very dry skin and 150 to the very hydrated skin [14]. Skin hydration was calculated by the following equation:

$$HR\% = \frac{100 \times M_p}{M_c}$$

where:

HR% = relative hydration;
Mp = average capacitance readings of product application areas;
Mc = average capacitance of the readings of the control region.

3.3.4. Skin pH Value Assessment

The study was conducted using equipment pHmeter Skin® PH 900 (Courage + Khazaka) and the measures were carried in delimited areas of volunteer's forearm skin.

3.3.5. Skin Oiliness Evaluation

The equipment used for the determination of skin oils was Sebumeter®, and measures were made in volunteer's forearm skin in delimited areas. The measuring time is 30 s for each point measured. The transparency of the plastic strip was evaluated, thereby quantifying the presence of lipid content on the skin surface, and the result is expressed in g of fatty material/cm² [15].

3.4. Statistical Analysis

The results were presented as mean ± standard deviation (Microsoft Office Excel 2007 software). The analysis of values' variance was made, considering a 95% significance level, and non-parametric ANOVA (Prism Software GraphPad® Prism version 4.00), followed by multiple comparisons by Newman–Keuls test.

4. Results

4.1. Formulation Studies

Nonionic surfactants are widely selected for cosmetic formulations, because of lower potential skin irritation [16]. Surfactant pairs composed by PEG castor oil 54 Ethylene Oxyde/sorbitan monooleate at $HLB_{required}$ values of 10.0 produce the most stable emulsions for babassu oil, as it was present in the right corner of ternary diagram (Figure 1A).

From Figure 1B, it was observed that the derivatives' formulas are around point 36, numbered 37 to 44, and it can be observed that the nanoemulsions were translucent, clear, and with an intense bluish reflection, without macroscopic signs of instability (Figure 1C,D). On the other hand, if the amount of oil incorporated was increased, and surfactant quantity was reduced in the formula, it was observed as a loss in bluish color reflection, and loss of translucency of the nanoemulsions.

Formula 38 had a bluish reflection and translucent appearance, and after centrifuge test and thermal stress analysis, was macroscopically stable.

Areca, licorice and portulaca extracts were added separately at the recommended concentration in scientific literature (3.0%). Macroscopic analysis was performed, employing 24 h old nanoemulsions. Stable babassu oil nanoemulsions were obtained with 3.0% w/w vegetable extract added separately, or blended in and now designated as F-38J, with HLB value = 10.0 (sorbitan monooleate/PEG 54 castor oil added with three vegetable extracts), and were chosen for comparative studies. Figure 2 shows size distribution for both studied F-38 and F-38J nanoemulsions.

In order to determine physical and chemical differences between F-38 and F-38J samples, they were assessed for particle size distribution, polydispersity index, pH, electrical conductivity and refractive index methods, before and after thermic stress (Table 4).

There was no statistically significant difference ($p > 0.05$) between F- 38 analyzed before and after the thermal stress test in all evaluated parameters. For F-38J nanoemulsions, it was observed that there is a statistically significant difference in polydispersity index and electrical conductivity values before and after the test.

Also, there was no statistically significant difference ($p > 0.05$) between the nanoemulsions F-38 analyzed before and after the freeze–defrost cycle test in the evaluated parameters (Table 4).

Babassu oil nanoemulsions, with and without extracts, were subjected to storage at temperatures of 4 ± 2 °C, 25 ± 5 °C and 45 ± 5 °C. The pH, electrical conductivity, refractive index values, particle size distribution, polydispersity index, and Ostwald ripening phenomenon were determined with samples that were 1, 7, 15, 30, 45, 60, 90, and 120 days old. For both F-38 and F-38F nanoemulsions stored at 4 ± 2 °C, 25 ± 5 °C, no statistically significant difference was observed from the first and after 120 days. Both formulas, when exposed at 45 ± 5 °C, showed a significant decrease in pH value after 120 days. For electrical conductivity, F-38J shows similar behavior as F-38: an increase in electrical conductivity values for samples stored at 45 ± 5 °C, while the samples stored at 4 ± 2 °C, and 25 ± 5 °C showed no statistically significant difference. These results indicate the presence of signs of instability in the system with increasing temperature (45 ± 5 °C). These results corroborate to the results of [3] in the development of passion fruit and lavender oil nanoemulsions.

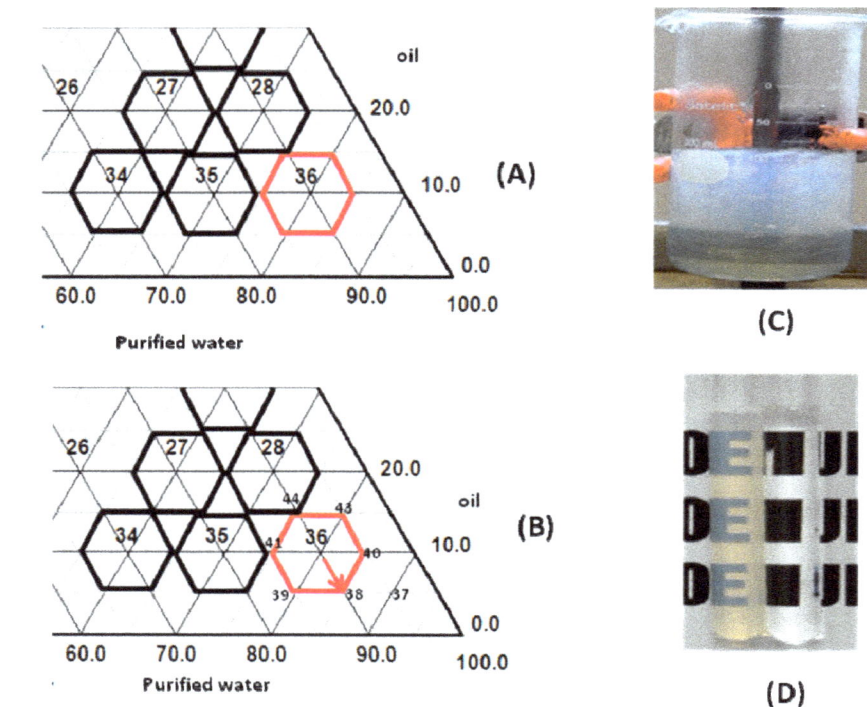

Figure 1. (**A**) Details of the phase diagram for babassu oil/water; (**B**) phase diagram—5.0% to 5.0%/surfactant (castor oil 54 EO/sorbitan monooleate, HLB value = 10.0); (**C**) nanoemulsion preparation; and (**D**) nanoemulsion.

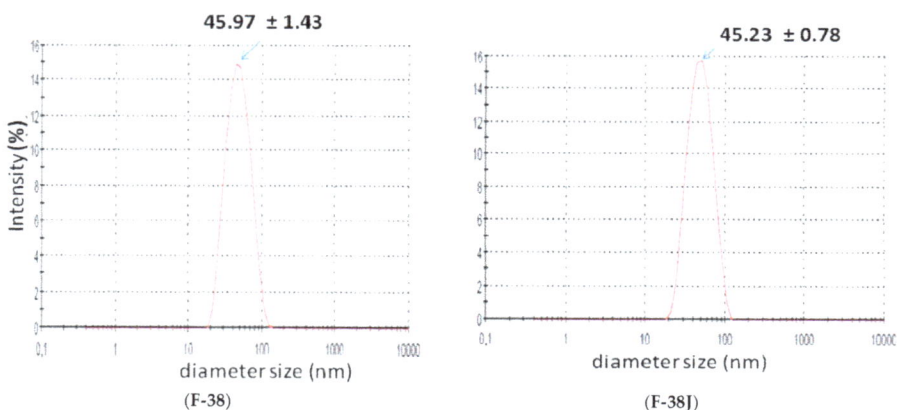

Figure 2. Particle size distribution comparison for babassu oil nanoemulsion without (**F-38**) and with (**F-38J**) vegetable extracts.

Table 4. Thermal stress and freeze–defrost analysis for F-38 and F-38J babassu nanoemulsions oil.

Thermal Stress				
Parameters	**F-38**		**F-38J**	
	Before	**After**	**Before**	**After**
droplets size (nm)	45.97 ± 1.43	45.77 ± 0.65	45.23 ± 0.38	57.83 ± 6.15
polydispersity index	0.106 ± 0.004	0.093 ± 0.020	0.091 ± 0.008	0.137 ± 0.022
pH value	6.32 ± 0.07	6.27 ± 0.11	6.53 ± 0.03	6.23 ± 0.05
electrical conductivity (μS/cm)	283.25 ± 2.56	290.19 ± 3.77	406.12 ± 5.91	453.14 ± 6.7
refractive index	1.359 ± 0.001	1.361 ± 0.001	1.362 ± 0.001	1.371 ± 0.004
Freeze–Defrost Cycles				
Parameters	**F-38**		**F-38J**	
	Before	**After**	**Before**	**After**
droplets size (nm)	45.97 ± 1.43	46.50 ± 0.60	45.23 ± 0.38	51.03 ± 477
polydispersity index	0.106 ± 0.004	0.103 ± 0.001	0.091 ± 0.008	0.125 ± 0.022
pH value	6.32 ± 0.07	6.42 ± 0.10	6.53 ± 0.03	6.45 ± 005
electrical conductivity (μS/cm)	283.25 ± 2.56	287.93 ± 3.98	406.12 ± 5.91	417.13 ± 11.19
refractive index	1.359 ± 0.001	1.359 ± 0.001	1.362 ± 0.001	1.363 ± 0.002

4.2. Nano Emulsion In Vitro Evaluation

4.2.1. Hen's Egg Test on the Chorioallantoic Membrane (HET-CAM)

The negative controls showed no change in the chorioallantoic membrane during the test. The positive control (SLS) was classified as severely irritant. It showed hyperemia and hemorrhage in under 30 s and 2 min, respectively, having a score of 10 on the scale (Table 5).

Table 5. Results for HET-CAM test.

O/W and Test Solutions	Time			Score	Classification
	Hyperemia	**Hemorrhage**	**Coagulation**		
F-38	≤30 s	____	____	5	slightly irritant
F-38J	≤30 s	____	____	5	slightly irritant
Surfactants solution	≤30 s	____	____	5	slightly irritant
Saline solution	____	____	____	0	practically not irritant
Purified water	____	____	____	0	practically not irritant
SLS solution	≤30 s	30 s < T ≤ 2 min.	____	10	irritant

In Table 5, we can see the occurrence of hyperemia for the nanoemulsions and also surfactants solution in a time less than 30 s, with no signs of hemorrhage and/or coagulation, corresponding to a score equal to 5, classifying the samples as slightly irritant. Then, both nanoemulsions F-38 and F-38J are safe, because the test did not show signs of hemorrhage and/or coagulation.

Our results are in accordance with Pereira [17] and Zanatta [18], in studies with nanoemulsions containing octyl methoxycinnamate, olive oil, grape seed oil, castor oil 30 EO, and sorbitan monooleate, which were considered slightly irritating, while the surfactants solution separately was moderately irritant. Zanatta [18] had employed buriti oil, sorbitan mono-oleate, and castor oil 40 EO.

Nanoemulsions containing surfactants derived from castor oil with different numbers of Polyoxyethylene (15, 30, 40 and 54 EO) and sorbitan monooleate were evaluated by the same method as Maruno [19], and the irritation potential was assessed as slightly irritant, reduced induced erythema, and demonstrated to be safe for cosmetics use.

These results indicate that in an emulsion, the surfactant is the most irritating component, and this characteristic increases with the number of ethylene oxides in the molecules.

4.2.2. Irritation Test in Red Blood Cell (RBC) System Cellular Model

This assay allows the quantification of adverse effects of isolated raw materials and finished products on the plasma membrane of red blood cells, and the consequent release of hemoglobin (hemolysis), quantified by denaturation of hemoglobin index, evaluating the hemoglobin oxidized form by spectrophotometry. The relationship between hemolysis and hemoglobin oxidation supplies an in vitro parameter, characterizing the effects of these substances [7,9].

In all nanoemulsion concentrations evaluated, lysis of red blood cells occurred, preventing the calculation of H_{50}, and consequently, the calculation of hemoglobin denaturation index, which enables the determination of the irritation degree of the emulsions. Then, no hemolytic activity was detected, and babassu oil nanoemulsions were classified as non-irritants, regardless of the addition or not of vegetable extracts.

Surfactant solution containing antimicrobial preservative and BHT presented hemolytic activity from 0.40 g/mL, indicating a dose-dependent activity. For the different tested concentrations, hemolytic activity was 0.99 ± 0.12 (0.40 g/mL), 6.03 ± 0.50 (0.50 g/mL), 12.09 ± 0.52 (0.60 g/mL), 52.76 ± 0.55 (0.70 g/mL), and $60.31 \pm 0.59\%$ (0.80 g/mL) (Figure 3).

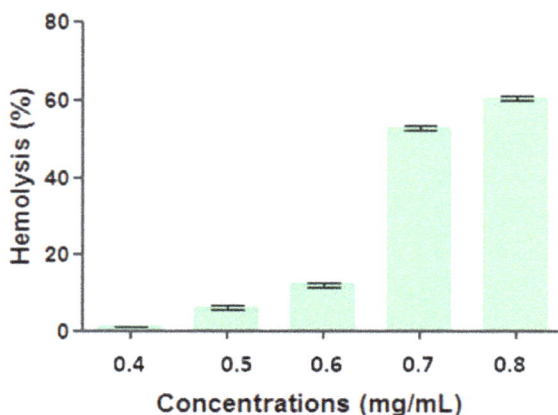

Figure 3. Hemolytic activity (%) versus surfactant solution concentration.

The formulations of babassu oil nanoemulsions evaluated by both HET-CAM and RBC method, were evaluated as slightly irritating, and non-irritating, respectively. Correlating the HET-CAM results, and those obtained by RBC, babassu oil nanoemulsions F-38 and F-38J may be indicated as safe for cosmetic use.

4.3. In Vivo Nanoemulsion Evaluation

4.3.1. Anti-Inflammatory Activity Evaluation by Edema Ear Rats Induced by Croton Oil

The ear edema induced by croton oil is a widely used model to assess the anti-inflammatory activity of steroidal and nonsteroidal drugs [10]. The main compound present in the croton oil (*Croton tiglium*) that acts as an inflammatory agent is 12-O-tetradecanoylphorbol-13-acetate (TPA). The application of the oil results in rapid accumulation of inflammatory cells, such as neutrophils and macrophages, to produce reactive oxygen species at the site [11].

After 4 hours of croton oil application in the ears of mice, a vivid hyperemia was visualized, and consequently, edema induction. For the ear thickness, the negative control already discounted from the left ear values presented 0.161 ± 0.022 mm of thickness (Figure 4). Groups II, III, IV, and V present a statistically significant difference ($p < 0.001$) compared to group I, with edema reduction of 25.97,

20.93, 44.67, and 66.93%. Groups II and III were the only groups that did not differ amongst the groups ($p > 0.05$).

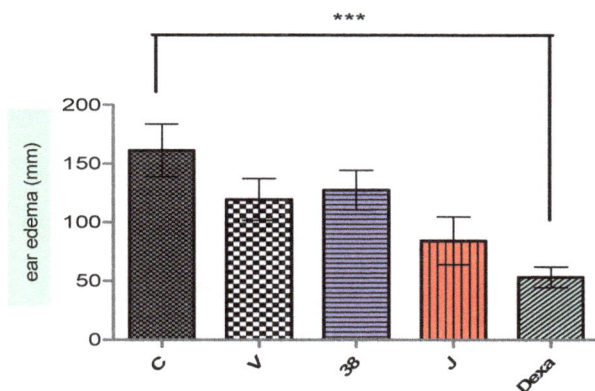

Figure 4. Ear edema induced by croton oil in untreated and treated animal groups. (**C**) Negative control; (**V**) vehicle; (**F-38**) nanoemulsion without extracts; (**F-38J**) nanoemulsion with extracts; (**Dexa**) dexamethasone solution. Values represent the mean and the respective standard deviations ($n = 6–8$). *** Significance level compared to the control group (**C**) ($p < 0.001$).

In relation to auricular mass, the negative control showed 5.9 ± 0.63 mg. Groups II, III, IV, and V have statistically significant differences compared to the negative control, with reduction of edema of 12.30, 16.09, 44.97, and 71.04%.

Thus, it is suggested that the chemical components in the F-38 and F-38J babassu oil nanoemulsions may be inhibiting the release of inflammatory mediators, or antagonizing the pharmacological receptors.

The results obtained by Sonneville-Aubrun et al. [20] in studies about the penetration rate of a nanoemulsion of 15.0% oil, in comparison to their corresponding macroemulsion, showed that the nanoemulsion penetrated significantly faster than the macroemulsion.

Nanoemulsions are an ideal system as vehicles for actives, even if they are insoluble in water. The small particle size allows for better adhesion of the active substance. In addition, spreadability, wetting, and penetration of the nanoemulsions are conferred by the low surface tension of the system [21].

4.3.2. Skin Hydration and Oiliness, and pH Evaluation

Hydration of Skin—Activity Evaluation

As we can see in Figure 5A, at 30 min F38J (with vegetable extracts) show the greater hydration values while the commercial product the smallest hydration power. The vegetable extracts collaborate to this fact and the F38 show the medium hydration value power. No significant statistical difference was observed between formulations hydration power at all analized times. The suggested nanoemulsions have major hydration values when compared to industrial product.

Commercial nanoemulsion comprises wetting agents and emollients including vegetable oil, glycerin, biosacharide gum, cetyl palmitate, silicone and sorbitol. However, the presence of several moisturizers compounds, do not provide significant higher hydration in relation to babassu oil nanoemulsions.

(A)

(B)

(C)

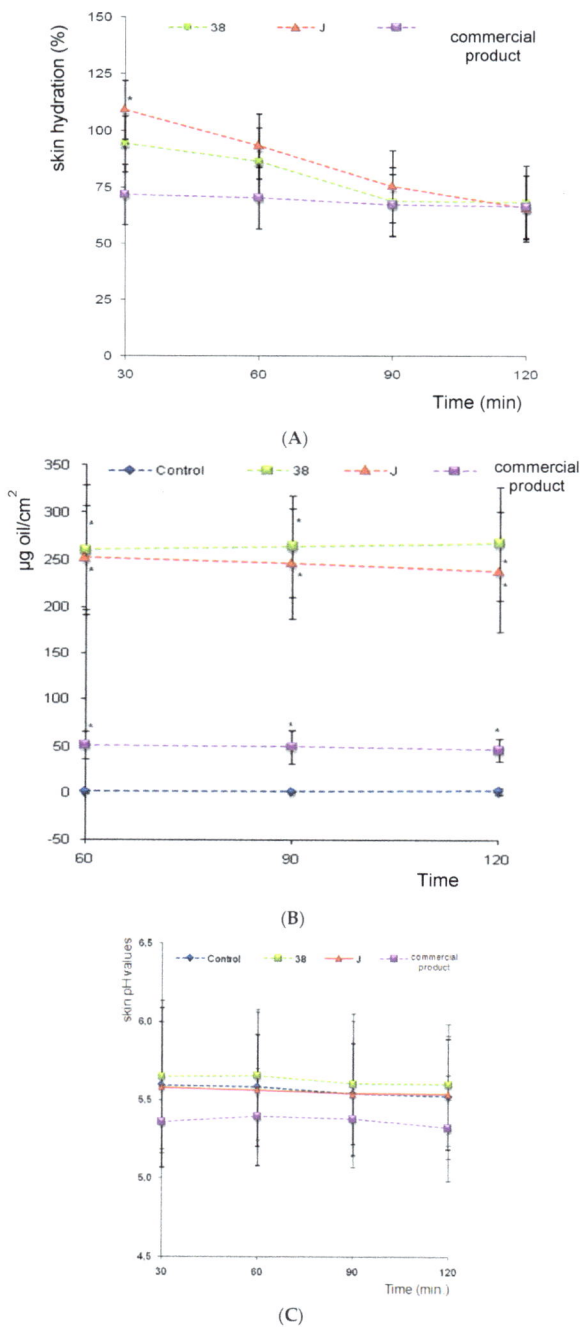

Figure 5. Hydration activity (**A**); skin oiliness (**B**); and skin pH evaluation (**C**) as a function of time. (*) no statistical differences between F38 and F38J. (*) Significance level compared to the control group (**C**) (*p* < 0.001).

Skin Oiliness Evaluation

The skin oiliness values for control regions did not change and were maintained around 0.0 to 1.0 mg/cm^2. Both F-38 and F-38J babassu oil nanoemulsions promoted an increase in the amount of skin oils, making statistically significant increases over the control area at 30, 60, 90, and 120 min. There was no statistical difference between F-38 and F-38J formulations, however, they diverge from the commercial nanoemulsions, which showed skin oiliness values about four times lower (Figure 5B).

Skin pH Evaluation

No statistically significant changes for nanoemulsions were observed in skin pH values, maintained at about 5.5, i.e., close to the normal pH value of the control area (4.2 to 5.9), indicating that the applied formulations were suitable for cosmetic use (Figure 5C).

5. Conclusions

Stable babassu oil nanoemulsion were obtained with surfactant pair fatty ester/PEG-54 castor oil with an HLB$_{required}$ value = 10.0 and having a droplet size of 45.97 ± 1.43 nm. The composition, order addition of components, and process parameters such as temperature, speed, and time of agitation, were critical in obtaining stable babassu oil nanoemulsions. There was no observed change in the size for both F-38 and F-38J nanoemulsions.

Although there are differences between the results of the HET-CAM and RBC tests, formulations F-38 and F-38J may be indicated for cosmetic use. In addition, both showed ear edema reduction capacity and skin moisturizing with no influence on skin pH value. Our results suggest the potential cosmetic use of babassu oil nanoemulsions.

Acknowledgments: This work had financial support from CAPES (Coordenação de Aperfeiçoamento de Pessoal de Nível Superior) and FAPESP (Fundação de Amparo a Pesquisa do Estado de São Paulo) under the protocol number (2009/07817-7; 2009/06152-1). We thank Lipo do Brasil, Croda and Oxiteno for supplying the raw material.

Author Contributions: V.C.G. was the student who carried out the laboratory work, analyzed the data and drafted the paper. O.F.S. contributed with the laboratory work; M.F. and M.M. contributed with the analysis of the data and the critical reading of the manuscript; P.A.R.-F. was the principal scientific supervisor of the study. He designed the study, supervised the laboratory work and contributed to critical reading of the manuscript.

Conflicts of Interest: The authors declare no conflict of interest.

Ethics Approval and Consent to Participate: All human experiments were implemented with approval of the Ethics Committee for Human Experiments (CEP/FCFRP n^r 204/2011). For HET-CAM analysis it's not necessary if the eggs have less than 15 days embryonated.

References

1. Pereira, T.A.; Guerreiro, C.M.; Maruno, M.; Ferrari, M.; Rocha-Filho, P.A. Exotic vegetable oils for cosmetic O/W nanoemulsions: In vivo evaluation. *Molecules* **2016**, *21*, 248. [CrossRef] [PubMed]
2. Rocha-Filho, P.A.; Attademo, M.; Oliveira, M.P.A.; Agostinho, L.C. Development of nanoemulsions with vegetal oils enhanced by melaleuca and lavander oils. In Proceedings of the International Symposium on Green Chemistry (ISGC), La Rochele, France, 16–19 May 2017.
3. Rocha-Filho, P.A.; Camargo, M.F.P.; Ferrari, M.; Maruno, M. Influence of lavander essential oil addition on passion fruit oil nanoemulsions: Stability and in vivo Study. *J. Nanomed. Nanotechnol.* **2014**, *5*. [CrossRef]
4. Davis, H.M. Analysis of creams and lotions. In *Newburger's Manual of Cosmetic Analysis*; Senzel, A.J., Ed.; Association of Official Analytical Chemists: Washington, DC, USA, 1997; Chapter 4; p. 32.
5. Prista, L.N.; Alves, A.C.; Morgado, R. *Tecnologia Farmacêutica*, 5th ed.; Fundação Calouste Gulbenkian: Lisboa, Portugal, 1996; Volume 1.
6. Formariz, T.P.; Urban, M.C.C.; Da Silva Júnior, A.A.; Gremião, M.P.D.; De Oliveira, A.G. Microemulsões e fases líquidas cristalinas como sistemas de liberação de fármacos. *Rev. Bras. Ciênc. Farm.* **2005**, *41*, 301–313. [CrossRef]

7. *Guia de Estabilidade de Produtos Cosméticos*; Agência Nacional de Vigilância Sanitária: Brasilia, Brazil, 2004; Volume 1. Available online: http://bvsms.saude.gov.br/bvs/publicacoes/cosmeticos.pdf (accessed on 7 September 2017).
8. Luepke, N.P.; Kemper, F.H. The HET-CAM test: An alternative to the draize eye test. *Food Chem. Toxicol.* **1986**, *24*, 495–496. [CrossRef]
9. Alves, E.N. Red Blood Cell (RBC)–Teste de Hemólise: Uma Alternativa ao Teste de Draize—Irritação Ocular na Avaliação do Poder Tóxico de Produtos Cosméticos No Controle de Qualidade. Mestrado Dissertação, Fundação Oswaldo Cruz, Instituto Nacional de Controle de Qualidade em Saúde, Rio de Janeiro, Brazil, 2003.
10. Tubaro, A.; Tragni, E.; Del Negro, P.; Galli, C.L.; Della Loggia, R. Anti-inflammatory activity of a polysaccharidic fraction of *Echinacea angustifolia*. *J. Pharm. Pharmacol.* **1987**, *39*, 567–569. [CrossRef] [PubMed]
11. Zanusso-Junior, G.; Melo, J.O.; Romero, A.L.; Dantas, J.A.; Caparroz-Assef, S.M.; Bersani-Amado, C.A.; Cuman, R.K.N. Avaliação da atividade antiinflamatória do coentro (*Coriandrum sativum* L.) em roedores. *Rev. Bras. Plantas Med.* **2011**, *13*, 17–23. [CrossRef]
12. Civille, G.V.; Dus, C.A. Evaluating tactile properties of skincare products: A descriptive analysis technique. *Cosmet. Toilet.* **1991**, *106*, 83.
13. Courage-Khazaka. Available online: http://www.cosmeticsonline.com.br/produtos/arquivos/A39_manual_mpa5_port_sebumeter_18-03-09.pdf (accessed on 28 June 2017).
14. Deccache, D.S. Formulação Dermocosmética Contendo DMAE Glicolato e Filtros Solares: Desenvolvimento de Metodologia Analítica, Estudo de Estabilidade e Ensaio de Biometria Cutânea. Master's Thesis, Faculdade de Farmácia, Universidade Federal do Rio de Janeiro, Rio de Janeiro, Brazil, 2006.
15. Manela-Azulay, M.; Cuzzi, T.; Araújo-Pinheiro, J.C.; Azulay, D.R.; Bottino-Rangel, G. Objective methods for analyzing outcomes in research studies on cosmetic dermatology. *An. Bras. Dermatol.* **2010**, *85*, 65–71. [CrossRef] [PubMed]
16. Bárány, E.; Lindberg, M.; Lodén, M. Unexpected skin barrier influence from nonionic emulsifiers. *Int. J. Pharm.* **2000**, *195*, 189–195. [CrossRef]
17. Pereira, G.G. Obtenção de Nanoemulsões O/A à Base de Óleo de Semente de Uva e Oliva Aditivadas de Metoxicinamato de Octila e Estudo do Potencial Antioxidante e Fotoprotetor das Emulsões. Mestrado Dissertação, Faculdade de Ciências Farmacêuticas de Ribeirão Preto, Universidade de São Paulo, Ribeirão Preto, Brazil, 2008.
18. Zanatta, C.F. Aplicação do Óleo de Buriti no Desenvolvimento e Emulsões e Estudo da Citotoxidade e Potencial Fotoprotetor em Cultivo Celular. Ph.D. Thesis, Faculdade de Ciências Farmacêuticas de Ribeirão Preto, Universidade de São Paulo, Ribeirão Preto, Brazil, 2008.
19. Maruno, M. Desenvolvimento de Nanoemulsões à Base de óleo de Gergelim Aditivadas de Óleo de Framboesa Para Queimaduras da Pele. Ph.D. Thesis, Faculdade de Ciências Farmacêuticas de Ribeirão Preto, Universidade de São Paulo, Ribeirão Preto, Brazil, 2009.
20. Sonneville-Aubrun, O.; Simonnet, J.T.; L'alloret, F. Nanoemulsions: A new vehicle for skincare products. *Adv. Colloid Interface Sci.* **2004**, *108–109*, 145–149. [CrossRef] [PubMed]
21. Wang, L.; Mutch, K.J.; Eastoe, J.; Heenan, R.K.; Dong, J. Nanoemulsions prepared by a two-step low-energy process. *Langmuir ACS J. Surfactants Colloids* **2009**, *24*, 6092–6099. [CrossRef] [PubMed]

cosmetics

MDPI

Article

Improving Skin Hydration and Age-related Symptoms by Oral Administration of Wheat Glucosylceramides and Digalactosyl Diglycerides: A Human Clinical Study

Valérie Bizot [1,*], Enza Cestone [2], Angela Michelotti [2] and Vincenzo Nobile [2]

[1]   Extraction Purification Innovation France–E.P.I. France, R&D department, 3 rue de Préaux, Villers sur Fère 02130, France

[2]   Farcoderm Srl, Member of Complife Group, Via Mons Angelini 21, San Martino Siccomario, Pavia 27028, Italy; enza.cestone@complifegroup.com (E.C.); angela.michelotti@complifegroup.com (A.M.); vincenzo.nobile@complifegroup.com (V.N.)

*   Correspondence: v.bizot@epifrance.fr; Tel.: +33-323-822897

Received: 13 June 2017; Accepted: 18 September 2017; Published: 21 September 2017

Abstract: Ceramides are known to play a key role in the skin's barrier function. An age-dependent decrease in ceramides content correlates with cutaneous clinical signs of dryness, loss of elasticity, and increased roughness. The present placebo-controlled clinical study aims to evaluate if an oral supplementation with glucosylceramides (GluCers) contained in a wheat polar lipids complex (WPLC) was able to improve such skin conditions. Sixty volunteers presenting dry and wrinkled skin were supplemented during 60 days with either a placebo or a WPLC extract in oil or powder form (1.7 mg GluCers and 11.5 mg of digalactosyldiglycerides (DGDG)). Skin parameters were evaluated at baseline and after 15, 30, and 60 days of supplementation. Oral intake of WPLC significantly increased skin hydration ($p < 0.001$), elasticity, and smoothness ($p < 0.001$), and decreased trans epidermal water loss (TEWL) ($p < 0.001$), roughness ($p < 0.001$), and wrinkledness ($p < 0.001$) in both WPLC groups compared to placebo. In both WPLC treated groups, all parameters were significantly improved in a time-dependent manner compared to baseline. In conclusion, this study demonstrates the positive effect of oral supplementation with GluCers on skin parameters and could reasonably reinforce the observations made on mice that orally-supplied sphingolipids can reach the skin.

Keywords: hydration; skin barrier function; anti-aging; glucosylceramides; food supplement; human clinical trial

---

1. Introduction

An important function of the skin is to provide an effective barrier against the loss of water and electrolytes. The properties of this permeability barrier are correlated with the role of *stratum corneum* (SC) lipid lamellae in forming the intercellular layer with a unique and very different composition from the lipids that compose biological membranes. Human SC lipids are composed of 50% ceramides, 25% cholesterol, and 15% free fatty acids (FFAs). All three components are required for skin integrity, especially the ceramides, which play a crucial role in bilayer system formation [1]. Ceramides form a complex and structurally heterogeneous group of sphingolipids; they are composed of a sphingoid base linked by an amide bond with a fatty acid (FA). Among them, acyl-ceramides (where linoleic acid is esterified on ω-hydroxy fatty acids) play an important role in the cohesion between the different intercellular lipid membranes and in water retention [2,3].

With aging, human skin becomes thinner, wrinkled, and loses some of its elasticity.

Dryness observed in aged skin correlates with an overall decrease of approximately 30% in total SC lipids [4]. In particular, it has been demonstrated that total SC ceramide content declines with age [5]. Additionally, it has been reported that age and season modify ceramides subclasses profiles in human SC. Indeed, the level of ceramide I and ceramide I linoleate (C18:2) decreases significantly between a young (26–29 years old) group and an older (57–60 years old) group [6,7].

Consequently, cosmetic research has focused for decades on finding evidence that topical application of ceramides and/or sphingolipids to the skin leads to improved moisture regulation, smoother and more elastic skin, and a general amelioration of the cutaneous barrier [8]. More recently, scientists have investigated the beneficial effects of oral supplementation with ceramides to improve dry skin, skin aspect, and associated discomforts. Animal studies have described ceramide bioavailability after ingestion. Dietary glycosylceramides were found to metabolize in rat small intestine: about half of the dietary substrate was absorbed, and was found in portal blood after hydrolysis by ceramidases in the gastrointestinal tract [9]. Even though a large proportion of ingested sphingolipids are excreted in the feces, animal studies reported that after oral intake radiolabeled ceramides are metabolized, absorbed and distributed to many tissues, including the skin. In rats, orally administered radiolabeled ceramides were shown to be delivered to the epidermis [10]. In another study on mice, after administration of ^{13}C-labeled dihydroceramides, their metabolite ^{13}C-labeled sphinganine was clearly detected in the skin, liver, skeletal muscle and synapse membrane in the brain [11]. It has also been shown that orally-administrated radiolabeled D2-sphingosine is transferred to the skin, from dermis to epidermis in an unchanged structural form, and further generates radiolabeled glucosylceramides and ceramides by in vivo biosynthesis in mice [12].

In addition, there is a slowly growing body of evidence from animal and human clinical studies that oral supplementation with ceramides may be beneficial for skin permeability barrier homeostasis [13–18] and parameters such as hydration and/or barrier function, elasticity, and recovery after induced disruption of barrier dysfunction.

The main dietary sphingolipid species that we ingest from cereals and plants is represented by glucosylceramides [12,19–24]. Cereal sphingolipids intake represents an average of 76 mg/day [24], corresponding to more than 25% of daily total sphingolipids.

Based on this interesting background, a proprietary wheat polar lipid complex (WPLC) was developed and produced in purified forms: a concentrated powder (WPLC-P) with high polar lipids content (≥98%) and an oil form (WPLC-O) rich in polar lipids (≥45%). It contains sphingolipids, also called phytoceramides, that included glucosylceramides and a wheat natural emulsifier, digalactosyl diglycerides (DGDG). In a pilot human clinical study, we have demonstrated that oral intake of 20 mg/day of WPLC in powder form (WPLC-P) can improve skin's moisturization index, elasticity, and skin microrelief significantly after only 15 days. Indeed, skin smoothness was increased while skin roughness and microwrinkles were decreased (unpublished data).

Based on these primary data, we decided to test the efficacy of oral intake of purified WPLC (containing glucosylceramides and DGDG) on skin hydration, barrier function, and aged-related symptoms in a placebo-controlled, randomized, and double-blind clinical trial with healthy human volunteers with dry and wrinkled skin.

2. Materials and Methods

2.1. Trial Design

This was a monocentric double-blind, placebo-controlled, randomized study performed on sixty healthy Caucasian female subjects. All of the study procedures were carried out according to the World Medical Association's (WMA) Helsinki Declaration and its amendments (Ethical Principles for Medical research Involving Human Subjects, adopted by the 18th WMA General Assembly Helsinki, Finland, June 1964 and amendments). The study protocol and the informed consent form were approved by an independent regional ethical committee (Code N 22-102009, Comitato Etico di Ricerca del Centro Analisi Monza S.p.A., Monza, Italy).

Information on the objectives and procedure of the trial, on dietary recommendations and on study benefits and risks were provided to the volunteers before their participation. The subjects were then asked for written informed consent before participating in the trial.

After a 5-day conditioning period, the volunteers were randomly assigned to three different intervention groups: placebo, WPLC-O (wheat polar lipids complex-oil form, 70 mg/day) and WPLC-P (wheat polar lipids complex-powder form, 30 mg/day). The conditioning period was used to allow all subjects to adapt their habits to the study requirements. During this period, all of the volunteers received a basal night and day cream (with no claimed effect on skin by topical application) in order to standardize topical application on the skin. The cream was applied by the volunteers to the face twice a day, in the morning and at night. Skin hydration was assessed at baseline and after 15, 30, and 60 days of supplementation with the tested products using Corneometer® CM 825 (Courage + Khazaka electronic GmbH, Cologne, Germany) and Tewameter® TM300 (Courage + Khazaka electronic GmbH, Cologne, Germany).

All of the analyses were centralized by Farcoderm Srl, Member of Complife Group, Via Mons Angelini 21, San Martino Siccomario 27028, Pavia, Italy.

No changes occurred in the methods after trial commencement.

2.2. Participants

2.2.1. Eligibility Criteria for Participants

Healthy women aged 30–60 years presenting dry skin (Corneometer® value < 50) and showing clinical signs of face skin aging related to photoaging (with medium photoaging signs, dry and devitalized skin, pale/greyish skin, early aging signs caused by slowing in cell activity) or mild-to moderate chrono-aging according to Fitzpatrick classification [25] were recruited. Corneometer® value for dry skin was set up based on "Information and Operating Instructions for the Corneometer® CM 825 Stand-alone" and with software version "CM825 alleine English 07/2007 DK" and on Farcoderm lab experience.

Exclusion criteria were an abnormal medical check-up, an obvious skin disease, an abnormal body weight (Body mass index: BMI < 19 and > 30 kg/m^2), a known history of lipid metabolism disorders, and intensively exposure to the sun (natural or artificial) for at least two months. Additional, exclusion criteria were pregnancy or intention to become pregnant, lactation, food allergy/intolerance, and participation in another similar study within the last two months prior to enrollment in the present study, using products containing moisturizing and/or anti-aging active ingredients taken orally or applied topically. The study further excluded subjects using tanning beds and undergoing dermatological (including peeling) or pharmacological (either local or systemic, i.e. corticoids, retinoids, vitamins) treatments within the last two months before the beginning of the study. Finally, all subjects were asked not to change their lifestyle, toilettes and dietary habits. Indeed, subjects unable/unwilling to comply with protocol requirements (including accordance not to use any other cosmetic product than the basal cream given at the beginning of the study) were not included in the study.

2.2.2. Settings and Locations

The study took place at Farcoderm Srl dermatological facilities in San Martino Siccomario (PV), Italy. Farcoderm Srl is an independent certified testing laboratory, collaborating with the University of Pavia, for in vitro and in vivo safety and efficacy assessment of cosmetics, food supplements, and medical devices.

2.3. Interventions

2.3.1. Experimental Wheat Polar Lipids Complex: Production and Characterization

A food-grade wheat polar lipids complex (WPLC) was extracted and purified according to a proprietary manufacturing process. Two grades were produced: an oil form (WPLC-O) and a concentrated powder form (WPLC-P). Briefly, for WPLC-O, this process consisted of successive water/ethanol extractions. After solid/liquid separation, the oil extract was concentrated under vacuum. WPLC-P was obtained by successive water/ethanol extractions, solid/liquid separation, and purification with acetone, followed by high-vacuum drying. Both forms of WPLC were extracted from selected wheat (*Triticum aestivum*, also named *vulgare* or *sativum*) endosperm flour.

WPLC was firstly characterized by mass spectrometry (MS). A direct electrospray/MS was performed for galactolipid characterization. Hydrolyzed WPLC was analyzed by gas chromatography/mass spectrometry (GC/MS) after derivatization and secondly by electrospray ionization/MS to assess the molecular mass of the substance and the filiation of fragments obtained by collision-activated dissociation (CAD) using an ESQUIRE (Bruker, Wissembourg, France) mass spectrometer with an ion trap. MS data on WPLC highlighted the presence of galactolipids, including digalactosyldiglycerides (DGDG) and glycosphingolipids, particularly glycosylceramides (Table 1).

Table 1. Lipid and fatty acids (FAs) composition identified and characterized by GC/MS.

Constituent Structures	Weighted Molecular Mass
Major GALACTOLIPIDS	
Digalactosyl diglycerides (DGDG) FAs composition: C18:2	940 g/mol
Major GLYCOSYLCERAMIDES	
Sphingoid bases: t 18:0 phytosphingosine (sphinganine) d 18:1 sphingosine (8 sphingenine) FAs composition: C16 to C18 (saturated and unsaturated) Sugar composition: Minimum 1 sugar	Average: 737 g/mol (from 716 to 756 g/mol)

Secondly, to obtain more details in WPLC sphingolipid characterization, an ultra-performant liquid chromatography–electrospray mass spectrometry/mass spectrometry (UPLC–ESI-MS/MS, Waters, Manchester, UK) study was performed using multiple reaction monitoring (MRM) mode, a method in which the eluate was continuously scanned for selected precursor–product ion pairs to enhance the sensitivity and specificity of the analysis of different classes of sphingolipids, as described in a recent publication [26]. Results were consistent with the first MS analysis, particularly for the identification of glycosylated phytoceramides. This UPLC–ESI-MS/MS study highlighted that WPLC contains more than 162 individual molecular species of ceramides, including glycosylceramides (data not shown). This high diversity in ceramides structure correlates with the presence of four sphingoid bases (t 18:0, t 18:1, d 18:0, d 18:1) and an amine linkage with a large diversity of FAs (from C16 to C26, in non-hydroxylated and hydroxylated forms). In the glycosphingolipids class, one species was clearly identified as glucosylceramide, where one molecule of glucose is esterified on the sphingoid bases of ceramides. At present, there is no routine technical solution to quantify all ceramide species. Thus, the more detailed analysis in WPLC has focused on the GluCers class, which have been previously defined as the principal markers of ceramides from plant sources [19–21]. Data are presented in Table 2.

Table 2. GluCers identified and characterized by UPLC-ESI-MS/MS.

GluCers Structure	GluCers Molecular Mass
Sphingoid bases: t 18:0 = phytosphingosine (4-hydroxysphinganine) t 18:1 = hydroxy sphingosine (4-hydroxy-8-sphingenine) d 18:0 = dihydrosphingosine (sphinganine) d 18:1 = sphingosine (8-sphingenine) FAs composition: C16 to C26 (hydroxylated and non-hydroxylated forms, saturated and unsaturated forms) Sugar composition: Monoglycosylated with 1 molecule of glucose	Average: 760 g/mol (from 716 to 872 g/mol)

High-field NMR spectra for proton (^{1}H) and carbon-13 (^{13}C) were obtained on a Bruker Avance DPX 500 spectrometer running at 500 MHz (Bruker, Wissembourg, France), after dissolution of WPLC in deuterated chloroform (CDCl3). The presence of ceramides is attested both by ^{1}H and ^{13}C-NMR (Figure 1).

Total esterified FA composition was quantified by GC following the standard methods references NF EN ISO 12966-2 and NF EN ISO 5508. Both forms of WPLC contained high levels of polyunsaturated fatty acids (>78%), including linoleic acid C18:2 (n−6) >60% for WPLC-O and >65% for WPLC-P.

Quantification of DGDG content was performed by an in-house validated HPLC method by reverse phase chromatography/ELSD using wheat DGDG as standard, from Larodan (ref: 59-1210-8).

Total sphingolipids content was quantified by an indirect internal method based on the following principle: one molecule of sphingolipids (or glycosphingolipids) contains one atom of nitrogen. Total nitrogen content was quantified by the Dumas method (NF EN ISO 16634-1). Nitrogen content coming from proteins and phospholipids was subtracted. Then, taking into account the average molecular weight of glucosylceramides characterized by MS (Table 1), calculation yielded an estimation of total sphingolipid content. More precisely, wheat GluCers markers in WPLC were quantified by normal phase high-performance liquid chromatography/evaporative light scattering detection (HPLC/ELSD) [22] using glucosylceramides from wheat as standard (Nacalaï ref NS170703).

Specifications of WPLC are reported in Table 3.

Table 3. WPLC active ingredients composition.

WPLC grade	DGDG (HPLC Method)	Total Sphingolipids (Nitrogen Method)
WPLC-P	≥40%	≥50%
WPLC-O	≥15%	≥15%

The content of wheat GluCers markers (HPLC) in the clinical tested batches of WPLC are as follows: 53 mg/g for WPLC-P and 24 mg/g for WPLC-O.

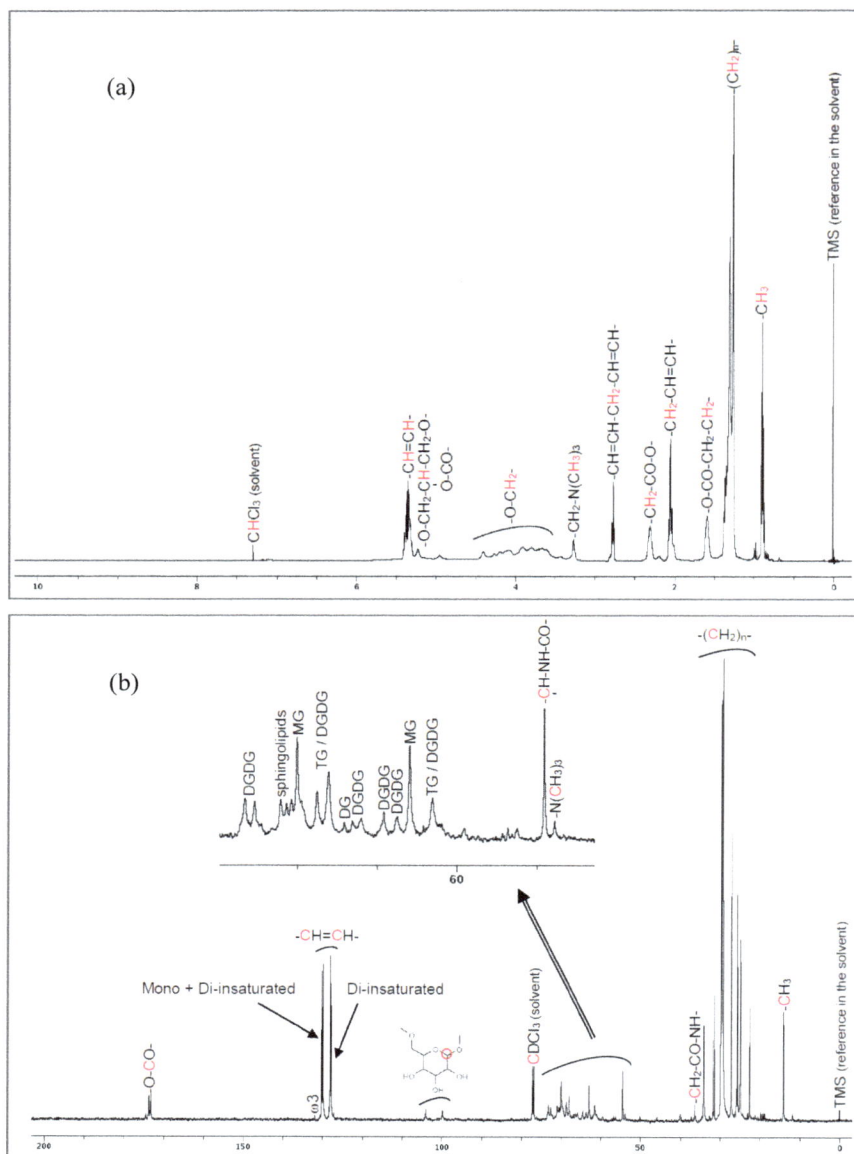

Figure 1. High field NMR spectra of WPLC; (**a**) ^{1}H-NMR spectra and (**b**) ^{13}C-NMR (TG: Triglycerides, DG: Diglycerides, MG: Monoglycerides).

2.3.2. Experimental Products: Description and Intake

During the 60-day experimental period, the daily oral intake was two capsules, at night before sleeping, containing either placebo (maltodextrin), WPLC-O extract (=70 mg/day) and WPLC-P extract (=30 mg/day). All capsules were identical in terms of size, color and odor and their composition is presented in Table 4. Indeed, WPLC-P and WPLC-O groups were, respectively, supplemented with 1.6 mg/day and 1.7 mg/day of wheat glucosylceramides and 12 mg/day and 11 mg/day of DGDG.

Table 4. Qualitative and quantitative formula of test products (per capsule).

Composition	Placebo (mg)	WPLC-O (mg)	WPLC-P (mg)
Maltodextrin	215	180	200
Dicalcium phosphate	100	100	100
Magnesium carbonate	50	50	50
Silice	30	30	30
WPLC-P	0	0	15
WPLC-O	0	35	0
Magnesium stearate	5	5	5
Total weight	400	400	400

The volunteers had to record their daily food consumption (using a 33 item food questionnaire) over the first 2 weeks of the intervention period (day 1 to 7 and then day 8 to 14) to ensure they did not change their dietary habits during the study course.

2.4. Outcomes

The primary efficacy endpoint was the evaluation of skin hydration measured in 5 replicates in the cheeks by Corneometer® CM 825 at day 60.

Secondary efficacy endpoints included the assessment of:

- skin hydration at day 15 and 30 by Corneometer® CM 825;
- transepidermal water loss (TEWL, 1 replicate measurement in the cheek, continuous measurement) measured by Tewameter® TM300 (Courage + Khazaka, electronic GmbH, Cologne, Germany) at D15, D30 and D60;
- skin elasticity (1 replicate measurement in the cheek) measured by Cutometer® MPA 580 (Courage + Khazaka, electronic GmbH, Cologne, Germany) (on time 2.0 s, off time 2.0 s, pressure 450 mbar, repetition 3, total time 12 s) at D15, D30 and D60;
- skin surface properties (smoothness, roughness and wrinkledness) (1 replicate measurement in the periocular area) measured by Visioscan® VC 98 (Courage + Khazaka, electronic GmbH, Cologne, Germany) at D15, D30 and D60;
- skin dermatological control by a board-certified dermatologist at D15, D30 and D60.

All of the measurements were carried out on cleansed faces under temperature- ($21 \pm 1\,^{\circ}C$) and humidity ($50 \pm 10\%$) controlled conditions. For 15 to 20 min before the beginning of the physical measurements, subjects were left to acclimate to the room conditions.

Volunteers' perception of product efficacy on skin following parameters: moisturizing level, skin elasticity, pulling sensation, desquamation, softness and smoothness of the skin evaluated by a self-assessment questionnaire at D60.

2.5. Sample Size

Sample size was calculated with a two-sided 5% significance level (α) and a power (1-β) of 80% taking into account the standard deviation and the variation obtained for skin hydration in a previous pilot study (data not shown). The statistical model was the one-sided one-sample *t*-test. According to the statistical model, a sample size of 13 subjects per group achieves 100% power to detect the variation of hydration seen in the pilot study. Seven additional volunteers were recruited in each group in case of possible dropout during the intervention period.

2.6. Randomization and Blindness

After the enrollment, subjects were randomly assigned to one of the three study groups, in a 1:1:1 ratio, to receive active products or a placebo. For allocation, a computer-generated (using PASS 11 statistical software, version 11.0.8 for Windows; PASS, LLC, Kaysville, UT, USA) restricted randomization list (Wei's urn algorithm) was used.

Subjects, investigator, and collaborators were kept blind to group assignment.

2.7. Statistical Methods

Statistical analysis was carried out on the intention-to-treat (ITT) population using NCSS 8 (version 8.0.4 for Windows; NCCS, LLC) running on a Windows 2008 R2 64 Edition server (Microsoft, Redmond, WA, USA). Data normality (both for raw data and variations vs. the baseline value) was verified using the Shapiro–Wilk W normality test and data shape. Comparisons were carried out to assess if the treatments (between-subjects) were source of variation of the measured endpoints. Data (differences adjusted for baseline) were submitted to RM-ANOVA followed by Tukey–Kramer Multiple-Comparisons post-test. The statistical significance was set at $p < 0.05$. p-values are reported as follows: ***, $p < 0.001$; **, $p < 0.01$; and *, $p < 0.05$. Results are reported as mean $\pm$ SD (Standard Deviation).

3. Results

3.1. Participant Flow and Recruitment

Subjects attended clinic visits at the time of randomization (baseline) and after 15, 30, and 60 days of product use. All of the randomized subjects (*n* = 20 per group) completed the study (Figure 2).

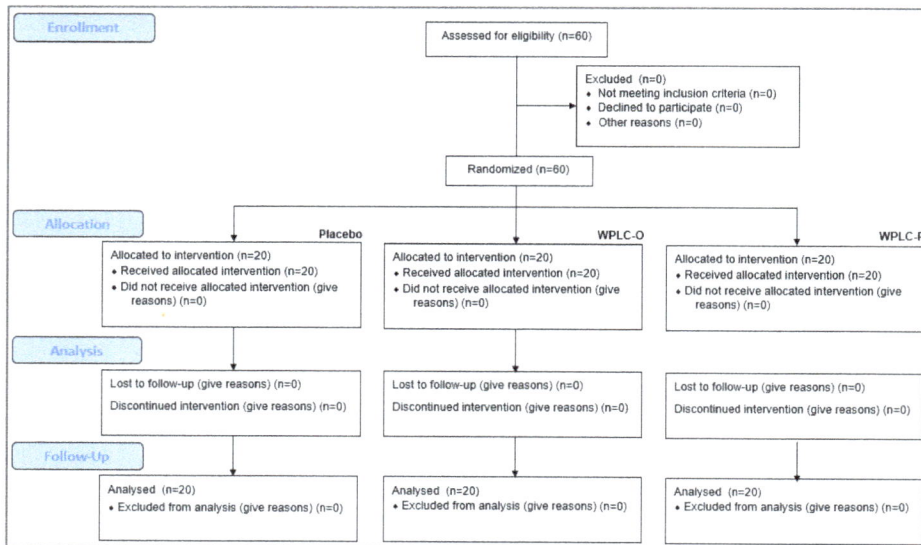

Figure 2. Study participant flow chart.

In accordance with the protocol requirement, the volunteers did not change their dietary habits during the study (data not shown).

The tested products were well tolerated overall by all the volunteers throughout the study. Self-assessment questionnaire results demonstrated that tolerance was found to be "fairly good" and "excellent" by 95%, 90% and 85% of the volunteers respectively supplemented with placebo, WPLC-P and WPLC-O products.

3.2. Baseline Data

Subjects' baseline demographics and clinical characteristics are presented in Table 5 for each group. During the baseline visit skin hydration was measured from Monday to Friday in order to obtain a consolidate value.

Table 5. Subjects' baseline demographic and clinical characteristics.

Baseline characteristics	Placebo	WPLC-O	WPLC-P
Number of subjects	20	20	20
ge (years)	48.3 ± 8.6	44.3 ± 8.7	45.7 ± 9.7
Weight(kg)	60.7 ± 10.1	59.0 ± 10.3	58.1 ± 6.6
BMI (kg/m^2)	22.9 ± 3.2	22.5 ± 3.2	22.2 ± 2.9
Menopause	8 (40%)	5 (25%)	6 (30%)
Smokers	5 (25%)	8 (40%)	8 (40%)
Cigarettes (n/day)	11.4 ± 7.5	12.4 ± 8.9	11.0 ± 6.6
Skin hydration (c.u)	44.5 ± 4.5	44.7 ± 5.0	43.3 ± 5.6
TEWL (g/h/m^2)	10.9 ± 2.7	11.3 ± 3.1	11.6 ± 3.3
Skin elasticity (Ratio Ua/Uf)	0.564 ± 0.064	0.581 ± 0.066	0.570 ± 0.074
Skin smoothness (SEsm (a.u.))	38.97 ± 5.88	38.86 ± 5.02	39.16 ± 5.61
Skin roughness(SEr (a.u.))	1.99 ± 0.37	1.90 ± 0.39	1.91 ± 0.36
Skin wrinkledness(SEw (a.u.))	35.37 ± 3.52	35.66 ± 3.98	34.89 ± 2.54

Data are averages ($\pm$standard deviation) or number of subjects (%).

3.3. Skin Hydration and Skin Barrier Function

Skin hydration values measured by Corneometer® are reported in Table 6.

Skin hydration was significantly increased ($p \leq 0.001$) after 15, 30 and 60 days of supplementation with WPLC-P and WPLC-O extracts compared to baseline (D0). On the contrary, in the placebo group, skin hydration was only increased after 30- and 60-day supplementation periods ($p < 0.05$ day 0). Variations in skin hydration measured at D15, D30 and D60 for both the WPLC-O and the WPLC-P groups were statistically different ($p < 0.001$) compared to placebo. There were no significant differences at each treatment time between the two active groups.

As reported in Table 6, similar improvements were obtained for TEWL. TEWL was significantly decreased ($p \leq 0.001$) after 15, 30 and 60 days of supplementation with WPLC-P and WPLC-O extracts compared to baseline. On the contrary, TEWL remained unchanged in the placebo group during the 60 days supplementation period ($p > 0.05$ vs. baseline and day 0). Variations in TEWL measured at D15, D30 and D60 for both the WPLC-O and the WPLC-P groups were statistically different ($p < 0.001$) compared to placebo. There were no significant differences at each treatment time between the two active groups.

Table 6. Skin Hydration and TEWL results after 15, 30 and 60 days of supplementation with WPLC-O and WPLC-P extracts measured by Corneometry® and Tewameter®. Data are averages ± SD in corneometric units (c.u.) and in TEWL (g/h/m²).

Time (days)	Corneometer® (c.u)			TEWL (g/h/m²)		
	Placebo	WPLC-O	WPLC-P	Placebo	WPLC-O	WPLC-P
D0	44.5 ± 4.5 [a]	44.7 ± 5.0 [a]	43.3 ± 5.6 [a]	10.9 ± 2.7 [b]	11.3 ± 3.1 [b]	11.6 ± 3.3 [b]
D15	46.2 ± 6.1 [a] (+3.6%)	54.8 ± 6.5 [b] (+23.4% [†])	53.6 ± 7.0 [b] (+24.9% [†])	11.2 ± 2.0 [b] (+5.3%)	9.4 ± 2.4 [a] (−16.2% [†])	9.7 ± 3.9 [a] (−16.0% [†])
D30	47.8 ± 5.8 [b] (+7.5%)	56.3 ± 6.0 [c] (+26.7% [†])	55.4 ± 5.8 [c] (+29.4% [†])	10.8 ± 2.8 [b] (+1.0%)	9.1 ± 2.5 [a] (−18.3% [†])	8.8 ± 2.9 [a] (−22.6% [†])
D60	48.1 ± 6.5 [b] (+8.2%)	60.9 ± 7.0 [d] (+37.2% [†])	58.6 ± 10.2 [d] (+36.8% [†])	10.9 ± 3.0 [b] (+3.8%)	9.0 ± 2.5 [a] (−19.5% [†])	9.3 ± 2.6 [a] (−19.0% [†])

In brackets is reported the % variation vs. D0; NS: Not significant; Significantly different from Day 0: a < b < c < d, $p < 0.05$; † Significantly different ($p < 0.001$) from the placebo group.

3.4. Skin Elasticity

Skin elasticity values measured by Cutometer® are reported in Table 7.

Table 7. Skin elasticity results after 15, 30 and 60 days of supplementation with WPLC-O and WPLC-P measured by Cutometer®. Data are averages ± SD in elasticity values R2 (Ua/Uf).

Time (days)	Skin elasticity (R2 (Ua/Uf))		
	Placebo	WPLC-O	WPLC-P
D0	0.564 ± 0.064 [a]	0.581 ± 0.066 [a]	0.570 ± 0.074 [a]
D15	0.558 ± 0.067 [a] (−0.7%)	0.665 ± 0.083 [b] (+14.9% [†])	0.661 ± 0.064 [b] (+16.6% [†])
D30	0.589 ± 0.083 [a] (+5.2%)	0.698 ± 0.072 [b] (+20.9% [†])	0.697 ± 0.060 [b] (+23.2% [†])
D60	0.606 ± 0.070 [a] (+8.3%)	0.788 ± 0.085 [c] (+36.6% [†])	0.766 ± 0.076 [c] (+35.9% [†])

In brackets is reported the % variation vs. D0; NS: Not significant; Significantly different from Day 0: a < b < c, $p < 0.05$; † Significantly different ($p < 0.001$) from the placebo group.

Skin elasticity was significantly increased ($p \leq 0.001$) after 15, 30 and 60 days of supplementation with WPLC-P and WPLC-O extracts compared to baseline (D0). On the contrary, it remained unchanged in the placebo group during the 60 days supplementation period ($p > 0.05$ vs baseline). Variation of skin elasticity in the WPLC-O and the WPLC-P groups was statistically different ($p < 0.001$) from the placebo group after 15, 30 and 60 days of supplementation. There were no significant differences at each treatment time between the two active groups.

3.5. Skin Aging Parameters

Smoothness, roughness and wrinkledness data are reported in Table 8 and in Table 9.

Skin roughness and wrinkledness were significantly decreased while smoothness was significantly increased after 15, 30 and 60 days of supplementation with WPLC-P and WPLC-O extracts compared to baseline (respectively $p \leq 0.001$, $p \leq 0.001$ and $p \leq 0.001$ in both groups). In the placebo group, skin roughness was significantly decreased after 60 days of supplementation while smoothness was significantly increased after 30 and 60 days of supplementation compared to baseline. However, variations in skin aging parameters measured at D15, D30 and D60 for both the WPLC-O and the WPLC-P groups were statistically different ($p < 0.001$, $p < 0.001$ and $p < 0.001$, respectively) compared to placebo. On the contrary, there were no significant differences at each treatment time between the two active groups.

Table 8. Skin smoothness and roughness values after 15, 30 and 60 days of supplementation with WPLC-O and WPLC-P extracts. Data are averages ± SD in SESm (a.u.) and SEr (a.u.) values.

Time (days)	Smoothness (a.u.)			Roughness (a.u.)		
	Placebo	WPLC-O	WPLC-P	Placebo	WPLC-O	WPLC-P
D0	38.97 ± 5.88 [a]	38.86 ± 5.02 [a]	39.16 ± 5.61 [a]	1.99 ± 0.37 [c]	1.90 ± 0.39 [c]	1.91 ± 0.36 [c]
D15	39.39 ± 5.40 [a] (+0.43%)	43.14 ± 4.18 [b] (+4.28% ∫)	43.48 ± 5.72 [b] (+4.32% ∫)	1.92 ± 0.62 [c] (−0.08%)	1.28 ± 0.51 [b] (−0.62% †)	1.32 ± 0.47 [b] (−0.59% †)
D30	40.87 ± 5.26 [b] (+1.90%)	48.33 ± 5.91 [c] (+9.47% †)	49.42 ± 5.58 [c] (+10.26% †)	1.77 ± 0.69 [c] (−0.22%)	0.87 ± 0.35 [a] (−1.03% †)	0.84 ± 0.40 [a] (−1.06% †)
D60	41.29 ± 5.54 [b] (+2.33%)	52.97 ± 5.11 [d] (+13.81% †)	52.33 ± 5.14 [c] (+13.18% †)	1.63 ± 0.61 [b] (−0.36%)	0.69 ± 0.28 [a] (−1.20% †)	0.68 ± 0.37 [a] (−1.23% †)

In brackets is reported the % variation vs. D0; NS: Not significant; Significantly different from Day 0: a < b < c < d, $p < 0.05$; † ($p < 0.001$) and ∫ ($p < 0.01$), significantly different from the placebo group.

Table 9. Skin wrinkledness variation results after 15, 30 and 60 days of supplementation with WPLC-O and WPLC-P. Data are mean ± SD in SEw (a.u.).

Time (days)	Skin Wrinkledness (SEw (a.u.))		
	Placebo	WPLC-O	WPLC-P
D0	35.37 ± 3.52 [d]	35.66 ± 3.98 [d]	34.89 ± 2.54 [d]
D15	36.11 ± 3.04 [d] (+0.74%)	32.56 ± 2.28 [c] (−3.10% [†])	32.14 ± 2.13 [c] (−2.75% [†])
D30	35.81 ± 3.36 [d] (+0.44%)	30.68 ± 2.20 [b] (−4.97% [†])	30.57 ± 2.50 [b] (−4.33% [†])
D60	35.01 ± 4.14 [d] (−0.36%)	28.87 ± 3.40 [a] (−6.79% [†])	28.60 ± 2.90 [a] (−6.29% [†])

In brackets is reported the % variation vs. D0; NS: Not significant; Significantly different from Day 0: a < b < c < d, $p < 0.05$; † ($p < 0.001$), significantly different from the placebo group.

3.6. Self-assessment Questionnaire Results

At day 60, self-assessment questionnaire results demonstrated that volunteers own perceptions of product efficacy are consistent with the measures performed on their skin. Results of the questionnaire are presented in Table 10.

Table 10. Results of the self-assessment questionnaire after 60 days of supplementation with WPLC-O and WPLC-P.

Questions and Self-Assessment	Placebo	WPLC-O	WPLC-P
How do you evaluate the food supplement efficacy concerning skin moisturizing level improvement?			
Insufficient	30%	5%	5%
Sufficient	50%	30%	35%
Fairly good	15%	45%	35%
Excellent	5%	20%	25%
How do you evaluate the food supplement efficacy concerning skin elasticity?			
Insufficient	45%	5%	5%
Sufficient	45%	30%	35%
Fairly good	5%	45%	35%
Excellent	5%	25%	25%
Do you think that your skin presents less "pulling sensation"?			
Yes	25%	80%	75%
No	75%	20%	25%
No opinion	0%	0%	0%
Do you think that your skin presents less desquamated?			
Yes	20%	85%	80%
No	75%	15%	15%
No opinion	5%	0%	5%
Do you think that your skin is softer and smoother?			
Yes	25%	85%	90%
No	65%	15%	10%
No opinion	10%	0%	0%
Do you think that the aspect of your skin looks better?			
Yes	15%	80%	80%
No	65%	15%	20%
No opinion	20%	5%	0%

4. Discussion

The purpose of this double-blind randomized controlled human clinical trial was to test, in a rigorous way, the hypothesis that even small amounts of sphingoid base derivatives, administered orally to humans, can afford measurable and significant and perceivable benefits to the skin, as suggested by the various animal studies and a few smaller-scale human trials [9–17] and our own preliminary study.

For this purpose, a specific purified wheat flour extract composed of wheat polar lipids rich in GluCers and DGDG was extracted, in both a powder and an oil based form, and formulated in capsules with identical aspect for oral intake over two months by healthy human volunteers presenting with dry skin and clinical skin aging signs. Placebo capsules were provided to a control group. Even if the level of active ingredient was similar, we tested both powder and oil form of WPLC to verify whether the galenic form had an impact or not on product efficacy.

In both treated groups, oral intake supplementation of 30 mg/day of WPLC-P or 70 mg/day of WPLC-O, providing an average of 1.7 mg of glucosylceramides and 11.5 mg of DGDG, induced a strong and highly significant improvement in skin hydration markers and skin anti-aging effects compared to the placebo group. As expected, given the almost identical content in the active components, DGDG and GluCers, there were no significant differences of efficacy between the two WPLC groups (powder form and oil form). Moreover, measures performed after 15, 30, and 60 days of supplementation showed that efficacy on each studied parameter increased significantly with time compared to the baseline. It is interesting to note that these significant effects appeared at the very early stage of the supplementation, in only 15 days and continued over the entire period of the study. With respect to the clinical pertinence of the statistically significant observation: while we did not include subjective "dermatological" scoring of the skin care effects, the self-assessment answers from the panelists in each group clearly showed strongly perceived benefits and therefore preferences for the verum capsules as opposed to the placebo. In view of the reported regulatory of food habits and intake, this factor cannot be considered as a source of data variation during the study period.

Like essential fatty acids, sphingosines and sphinganines, the sphingoid base structures in SC ceramides, cannot be synthetized de novo by the human organism. They are generated by degradation of extracutaneous lipids coming from the general diet [9,10,12].

The fate of ingested sphingolipids in human beings remains to be understood by further studies. Based on the present state of knowledge, we can reasonably assume that sphingolipid metabolism in humans will follow a similar path of digestion and absorption to that described in animal studies, producing sphingolipid metabolites that are absorbed intestinally and distributed to blood. Based on this hypothesis, metabolites of dietary sphingolipids, including ceramides and glucosylceramides, could reach the skin, enter the metabolic processes in the epidermis through the Golgi apparatus and lamellar bodies, and thus participate in the de novo synthesis of ceramides and other physiological skin lipids in situ. Consequently, orally taken sphingolipids could play their functional role in the stratum corneum to maintain or restore dry skin conditions and improve skin microrelief.

Cereals, particularly barley, corn, and wheat, are DGDG-rich [27]. In rats orally fed with DGDG, the fatty acids generated after hydrolysis were shown to be reesterified into triglycerides and phospholipids in the enterocytes and then transported by triglyceride-rich lipoproteins [28].

The presence of DGDG in WPLC is likely to improve the efficacy of GluCers by increasing intestinal absorption. For instance, as shown in a human clinical study [29], oral supplementation of phytosterol dispersed in a diacylglycerol-rich oil increased the health benefits of phytosterols compared to a triacylglycerol oil. The authors explained that the reason for this could be that diacylglycerol-rich oil is a better solvent for hydrophobic components as phytosterols. More recently, it was demonstrated that bioaccessibility of beta-carotenoids was markedly enhanced when solubilized in diacylglycerol (dioleate) before digestion compared to beta-carotene alone. The low bioavailability of lipophilic micronutrients is mainly caused by their limited solubilization into aqueous micelles, which hinders their ability to be taken up by the intestine [30].

DGDG is in fact a specific diacylglycerol where two galactoses are esterified, with high emulsifying properties [31]. By analogy with diacyl glycerol oil, DGDG could be largely responsible for an increase in the distribution of dietary sphingolipids including GluCERs in bile acid micelles after ingestion, followed by improvement of intestinal absorption of both intact forms and metabolites.

Thanks to the amphiphilic structure of DGDG, sphingolipids, including GluCERs that composed WPLC, can be dispersed in highly stable micellar dispersion in water.

Taking all these observations and the present study data into account, it appears reasonable to correlate the demonstrated improvements in age-related symptoms such as skin hydration, elasticity and smoothness with the single-parameter of WPLC supplementation in this human clinical trial.

5. Conclusions

In a human clinical trial based on a gold standard study design, we show that daily oral supplementation of purified wheat glucosylceramides and DGDG induced a strong and highly significant improvement in skin hydration markers compared to placebo, after only 15 days and beyond.

The level of efficacy on skin hydration and TEWL was higher and was detected faster than data reported in other human clinical studies on sphingolipids or ceramides oral supplementations. For the first time, we demonstrated the anti-aging properties induced by wheat sphingolipids, including GluCers and DGDG, issued from WPLC oral supplementation, on four skin markers concomitantly: elasticity, smoothness, roughness, and wrinkledness. Purified WPLC can be considered as an interesting food supplement ingredient inducing healthy and young-looking skin.

Although the clinical research has reached its aim, there were some unavoidable limitations. Indeed, interactions between the skin and the basal cream were possible and cannot be rejected a priori. Some of the possible interaction also takes into account the "placebo" effect. However, even if some interactions occurred, we think that they do not represent a limitation of study results interpretation. In fact, the level of skin parameter improvements obtained in the active treatment arms is significantly higher than the one measured in the placebo arm. In addition, some inflation of the overall alpha-error can occur due to the number of statistical tests. Further studies to investigate the bioavailability of wheat sphingolipids and DGDG metabolites after digestion, their effectiveness on ceramides metabolism pathway in the skin, the enhancement of ceramides content at the skin level, and the specificity of ceramide structures and associated lipids on their efficacy are planned.

Acknowledgments: We sincerely express our gratitude for valuable advice and comments to Claire Notin from SEPPIC SA, 22 Terrasse Bellini, Paris la Défense, Puteaux, France.

Author Contributions: V.M.B. conceived and designed the experiments; E.C. and A.M. performed the experiments; V.N. analyzed the data; V.M.B. wrote the paper.

Conflicts of Interest: The authors declare no conflict of interest.

References

1. Feingold, K.R.; Elias, P.M. Role of lipids in the formation and maintenance of the cutaneous permeability barrier. *Biochim. Biophys. Acta* **2014**, *1841*, 280–294. [CrossRef] [PubMed]
2. Jungersted, J.M.; Hellgren, L.I.; Jemec, G.B.; Agner, T. Lipids and skin barrier function–a clinical perspective. *Contact Dermatitis* **2008**, *58*, 255–262. [CrossRef] [PubMed]
3. Rabionet, M.; Gorgas, K.; Sandhoff, R. Ceramide synthesis in the epidermis. *Biochim. Biophys. Acta* **2014**, *1841*, 422–434. [CrossRef] [PubMed]
4. Ghadially, R.; Brown, B.E.; Sequeira-Martin, S.M.; Feingold, K.R.; Elias, P.M. The aged epidermal permeability barrier. Structural, functional, and lipid biochemical abnormalities in humans and a senescent murine model. *J. Clin. Investig.* **1995**, *95*, 2281–2290. [CrossRef] [PubMed]
5. Imokawa, G.; Abe, A.; Jin, K.; Higaki, Y.; Kawashima, M.; Hidano, A. Decreased level of ceramides in stratum corneum of atopic dermatitis: An etiologic factor in atopic dry skin? *J. Investig. Dermatol.* **1991**, *96*, 523–526. [CrossRef] [PubMed]
6. Rogers, J.; Harding, C.; Mayo, A.; Banks, J.; Rawlings, A. Stratum corneum lipids: The effect of ageing and the seasons. *Arch. Dermatol. Res.* **1996**, *288*, 765–770. [CrossRef] [PubMed]
7. Pons-Guiraud, A. Dry skin in dermatology: A complex physiopathology. *J. Eur. Acad. Dermatol. Venereol.* **2007**, *21*, 1–4. [CrossRef] [PubMed]

8. Lintner, K.; Mondon, P.; Girard, F.; Gibaud, C. The effect of a synthetic ceramide-2 on transepidermal water loss after stripping or sodium lauryl sulfate treatment: An in vivo study. *Int. J. Cosmet. Sci.* **1997**, *19*, 15–26. [CrossRef] [PubMed]

9. Nilsson, A.; Duan, R.D. Alkaline sphingomyelinases and ceramidases of the gastrointestinal tract. *Chem. Phys. Lipids* **1999**, *102*, 97–105. [CrossRef]

10. Ueda, O.; Hasegawa, M.; Kitamura, S. Distribution in skin of ceramide after oral administration to rats. *Drug Metab. Pharmacokinet.* **2009**, *24*, 180–184. [CrossRef] [PubMed]

11. Fukami, H.; Tachimoto, H.; Kishi, M.; Kaga, T.; Waki, H.; Iwamoto, M.; Tanaka, Y. Preparation of (13) C-labeled ceramide by acetic acid bacteria and its incorporation in mice. *J. Lipid Res.* **2010**, *51*, 3389–3395. [CrossRef] [PubMed]

12. Ueda, O.; Uchiyama, T.; Nakashima, M. Distribution and metabolism of sphingosine in skin after oral administration to mice. *Drug Metab. Pharmacokinet.* **2010**, *25*, 456–465. [CrossRef] [PubMed]

13. Kajimoto, O. Clinical investigation of skin-beautifying effect of a beauty supplement containing rice-derived ceramide. *J. New Rem. Clin.* **2002**, *51*, 62–72.

14. Uchiyama, T.; Nakano, Y.; Ueda, O.; Mori, H.; Nakashima, M.; Noda, A.; Ishizaki, C.; Mizoguchi, M. Oral intake of glucosylceramide improves relatively higher level of transepidermal water loss in mice and healthy human subjects. *J. Health Sci.* **2008**, *54*, 559–566. [CrossRef]

15. Asai, S.; Miyachi, H. Evaluation of skin-moisturizing effects of oral or percutaneous use of plant ceramides. *Rinsho Byori* **2007**, *55*, 209–215. [PubMed]

16. Hori, M.; Kishimoto, S.; Tezuka, Y.; Nishigori, H.; Nomoto, K.; Hamada, U.; Yonei, Y. Double-blind study on effects of glucosyl ceramide in beet extract on skin elasticity and fibronectin production in human dermal fibroblasts. *Anti-Aging Med.* **2010**, *7*, 129–142. [CrossRef]

17. Boisnic, S.; Branchet, M.C. Intérêt clinique d'un ingrédient alimentaire à visée hydratante: lipowheat™, étude randomisée en double aveugle versus *placebo*. *J. Med. Esthet. Chir. Dermatol.* **2007**, *34*, 239–242.

18. Guillou, S.; Ghabri, S.; Jannot, C.; Gaillard, E.; Lamour, I.; Boisnic, S. The moisturizing effect of a wheat extract food supplement on women's skin: A randomized, double-blind placebo-controlled trial. *Int. J. Cosmet. Sci.* **2011**, *33*, 138–143. [CrossRef] [PubMed]

19. Vesper, H.; Schmelz, E.M.; Nikolova-Karakashian, M.N.; Dillehay, D.L.; Lynch, D.V.; Merrill, A.H. Sphingolipids in food and the emerging importance of sphingolipids to nutrition. *J. Nutr.* **1999**, *129*, 1239–1250. [PubMed]

20. Yunoki, K.; Ogawa, T.; Ono., J.; Miyashita, R.; Aida, K.; Oda, Y.; Ohnishi, M. Analysis of sphingolipid classes and their contents in meals. *Biosci. Biotechnol. Biochem.* **2008**, *72*, 222–225. [CrossRef] [PubMed]

21. Sugawara, T.; Aida, K.; Duan, J.; Hirata, T. Analysis of glucosylceramides from various sources by liquid chromatography-ion trap mass spectrometry. *J. Oleo Sci.* **2010**, *59*, 387–394. [CrossRef] [PubMed]

22. Yan, X.; Vredeveld, D.J.; Missler, S.R. Quantification of glucosylceramides in wheat extracts using high-performance liquid chromatography with evaporative light-scattering detection. *Am. J. Food Sci. Nutr.* **2015**, *2*, 95–100.

23. Hellgren, L. Sphingolipids in the human diet-occurrence and physiological effects. *Lipid Technol.* **2002**, *14*, 129–133.

24. Wehrmuller, K. Occurrence and biological properties of sphingolipids—A review. *Curr. Nutr. Food Sci.* **2007**, *3*, 161–173. [CrossRef]

25. Fitzpatrick, R.E.; Goldman, M.P.; Satur, N.M.; Tope, W.D. Pulsed carbon dioxide laser resurfacing of photo-aged facial skin. *Arch. Dermatol.* **1996**, *132*, 395–402. [CrossRef] [PubMed]

26. Tellier, F.; Maia-Grondard, A.; Schmitz-Afonso, I.; Faure, J.D. Comparative plant sphingolipidomic reveals specific lipids in seeds and oil. *Phytochemistry* **2014**, *103*, 50–58. [CrossRef] [PubMed]

27. Sugawara, T.; Miyazawa, T. Separation and determination of glycolipids from edible plant sources by high-performance liquid chromatography and evaporative light-scattering detection. *Lipids* **1999**, *34*, 1231–1237. [CrossRef] [PubMed]

28. Ohlsson, L.; Blom, M.; Bohlinder, K.; Carlsson, A.; Nilsson, A. Orally fed digalactosyldiacylglycerol is degraded during absorption in intact and lymphatic duct cannulated rats. *J. Nutr.* **1998**, *128*, 239–245. [PubMed]

29. Meguro, S.; Higashi, K.; Hase, T.; Honda, Y. Solubilization of phytosterols in diacylglycerol versus triacylglycerol improves the serum cholesterol-lowering effect. *Eur. J. Clin. Nutr.* **2001**, *55*, 513–517. [CrossRef] [PubMed]

30. Nagao, A.; Kotake-Nara, E.; Hase, M. Effects of fats and oils on the bioaccessibility of carotenoids and vitamin E in vegetables. *Biosci. Biotechnol. Biochem.* **2013**, *77*, 1055–1060. [CrossRef] [PubMed]

31. Hauksson, J.B.; Bergqvist, M.H.J.; Rilfors, L. Structure of digalactosyldiacylglycerol from oats. *Chem. Phys. Lipids* **1995**, *78*, 97–102. [CrossRef]

cosmetics

MDPI

Article

Non-Targeted Secondary Metabolite Profile Study for Deciphering the Cosmeceutical Potential of Red Marine Macro Alga *Jania rubens*—An LCMS-Based Approach

Dhara Dixit [1,*] and C. R. K. Reddy [2]

[1] Department of Earth & Environmental Science, Krantiguru Shyamji Krishna Verma Kachchh University, Bhuj-Kachchh 370001, Gujarat, India
[2] Discipline of Marine Biotechnology & Ecology, CSIR-Central Salt and Marine Chemicals Research Institute, Gijubhai Badheka Marg, Bhavnagar 364002, Gujarat, India; crk@csmcri.res.in
* Correspondence: dharadixit2008@gmail.com; Tel.: +91-99-7909-0900

Received: 13 September 2017; Accepted: 23 October 2017; Published: 30 October 2017

Abstract: This study aims to unveil the cosmeceutical traits of *Jania rubens* by highlighting its mineral composition, antioxidant potential, and presence of bioactive molecules using non-targeted metabolite profiling. This study showed that among minerals, (macro), Ca (14790.33 + 1.46 mg/100 g dry weight (DW)) and in (micro) Fe (84.93 + 0.89 mg/100 g DW) was the highest. A total of 23 putative metabolites in the +ESI (Electrospray Ionization) mode of LCMS-TOF (Liquid Chromatography Mass Spectrometry-Time of Flight) were detected. Two anthocyanins—malonylshisonin and 4'''-demalonylsalvianin (m/z 825.19; anti-aging, antioxidant, anticancer properties) were detected. Two flavonoids, viz, medicocarpin and agecorynin C, 4'-O- methylglucoliquiritigenin—a flavonoid-7-O-glycoside, and 5,6,7,8,3',4',5'-heptamethoxyflavone, a polymethoxygenated flavone (m/z 415.15), were detected. Maclurin 3-C-(2'',3'',6''-trigalloylglucoside) (m/z 863.15) (antioxidant, antimicrobial and anticancer traits) and theaflavonin (m/z 919.18), belonging to the class of theaflavins (whitening and anti-wrinkle agent), were obtained. Pharmacologically active metabolites like berberrubin (m/z 305.1; antitumor activity), icaceine (m/z 358.24; anticonvulsant properties), agnuside (m/z 449.15; constituent for treatment of premenstrual syndrome), γ-coniceine (m/z 108.12; formulations to treat breast cancer), eremopetasitenin B2, and eremosulphoxinolide A (m/z 447.18; therapeutic effect of allergy and asthma) were observed. 6-O-Methylarmillaridin (m/z 445.18) (antimicrobial and antifungal) and simmondsin 2-ferulate, (m/z 534.21) (insecticidal, antifungal and antifeedant) were detected. Aromatic lignans, viz, 8-Acetoxy-4'-methoxypinoresinol, sesartemin, and cubebinone (m/z 413.16), in addition to an aromatic terpene glycoside, tsangane L3 glucoside (m/z 357.23), were detected. Zizybeoside I, benzyl gentiobioside, and trichocarposide were also detected. The determination of antioxidant potential was performed through assays such as like DPPH (2,2-diphenyl-1-picrylhydrazyl), FRAP (Ferric Ion Reducing Antioxidant Power), ABTS (2,2'-azino-bis(3-ethylbenz-thiazoline-6-sulfonic acid)), and total antioxidants. Therefore, this study progresses the probability for the inclusion of *J. rubens* as an ingredient in modern day cosmetic formulations.

Keywords: cosmeceutical; *Jania rubens*; non-targeted; metabolites; antioxidant

---

1. Introduction

"Cosmeceuticals" is a term derived from cosmetics and pharmaceuticals, indicating that a specific product contains certain active ingredients [1] with drug-like benefits that enhance or protect the appearance of human body [2]. Cosmeceuticals have medicinal benefits that affect the biological functioning of skin, depending upon the type of functional ingredients they contain. Marine macro algae have gained much attention nowadays in the context of cosmeceutical product development, owing to their remarkably rich bioactive composites. Bioactive components from such macro algal origins exhibit varied functional roles such as secondary metabolites, and these properties can thus be harnessed for the development of cosmeceuticals [3]. These secondary metabolites include polyphenols, alkaloids, cyclic peptide, phlorotannins, diterpenoids, polyketides, sterols, quinones, glycerols and polysaccharides [4]. They are not directly involved in any primary functions viz; growth and development, unlike the primary metabolites (amino acids, organic acids, lipids and other compounds). Instead, they play precise roles in providing protection against infections caused by ultraviolet radiation and pathogens [5]. Their other roles include pigmentation and reproduction [6]. Secondary metabolites can be distinguished based on their chemical structure, precursor molecules or synthetic pathways. A complete set of both primary as well as the secondary metabolites comprise the metabolome. Metabolomics is thus the understanding of such complex metabolic networks, as represented by the identification and quantification of all available metabolites of the corresponding biological system [7].

Skin ageing and wrinkle formation can also be caused by reactive oxygen species (ROS) which are produced due to oxidative stress [8]. Herbal cosmeceutical preparations are quite popular among customers, as these agents are mostly perceived by customers as safe, non-toxic, and exhibiting strong antioxidant activity. Such antioxidants act as guardians to counteract oxidative stress generated by excess ROS formation. Phenol and flavonoids in particular, are important classes of natural antioxidants. Consequently, there is a rise in exploring unique and effective antioxidants of botanical origin that can quench free radicals and ROS, in order to defend the skin from oxidative damages.

Currently, several studies have opened up insights emphasizing upon the role of biological activities of marine algae in promoting health, skin and beauty products. Until now, a few marine organisms have been explored for screening of some cosmeceutical compounds. For instance, *Ecklonia cava*, a phaeophyte, has already proven itself to be an interesting candidate based on its phlorotannin (eckol and dieckol) contents [9,10]. Apparently, an aqueous extract of another phaeophyte, *Macrocystis pyrifera*, has been found to foster the synthesis of a major component of the extracellular matrix of the skin called the hyaluronic acid [11]. Another report stated that the methanolic extract of a red alga named *Corallina pilulifera*, lowered the manifestation of MMP-2 and -9 (induced by UV radiations) in dermal fibroblast of humans [12]. An edible brown alga has been reported to cause a substantial reduction of cutaneous progerin, resulting in the stimulation of collagen production via its lipophilic extract [13]. Several other species like *Schizymenia dubyi*, *Endarachne binghamiae*, *Sargassum siliquastrum* and *Ecklonia cava* have already been documented to be virtuous natural alternatives for skin lightening. Furthermore, the red alga *Corallina ellongata* is used as a source for the extraction of phycoerythrin and certain other proteins, which have gained significance in therapeutics, immunodiagnostic as well as cosmetics [14].

Jania rubens, a benthic red marine macro alga belonging to the Corallinaceae family, possesses unique structural and functional features. It is described as an alga that is heavily infused with biochemically precipitated $CaCO_3$, in the form of calcite, through the phenomenon of biomineralization. A strong antifouling activity against mussels has been reported from this alga, strengthening the knowledge of a chemical defense mechanism of this alga [15]. *J. rubens* has been reported to protect itself against epibionts using a physical antifouling phenomena without the release of any chemicals bearing antifouling properties [16]. It has been studied for its antimicrobial potential [17] cytotoxic activity [18] and its antibacterial properties against human pathogens [19]. *J. rubens* (L.) Lamx has also been a source for the isolation of seven brominated diterpenes of the parguerene and isoparguerene series having marked antitumor activity [20]. It has also been a source for the isolation of xylogalactans, which have industrially applications as thickeners and gelling agents [21]. A *Jania rubens* extract is being proposed for formulating slimming cosmetics, as it apparently promotes the elimination of fats and the synthesis of collagen for smoothing out cellulite. This claim however, is not substantiated by the scientific literature. In addition to this, *J. rubens* extract is also being proposed for skin conditioning formulations. But to the best of our knowledge, this is the first comprehensive scientific study to assess its cosmeceutical virtues. Unveiling the secondary metabolite profile of such a resourceful marine algae would be quite interesting, as it may open new avenues for designing algal bio refinery for obtaining high value added products useful for future cosmetic applications. Therefore, the present study aims at undertaking a comprehensive study of *J. rubens* from the perspective of decoding its secondary metabolite framework, antioxidant potential, and biomineralization attributes for its possible inclusion as a cosmeceutical ingredient in future formulations.

2. Materials and Methods

2.1. Sample Collection

Jania rubens was collected from the Pingleshwar region along the Kachchh coast (23°54′ N, 68°48′ E), in Gujarat, India. Manual harvesting of the algal thalli was undertaken during the month of December, and the collected algal biomass was identified and confirmed by the experts at CSIR-CSMCRI (Council of Scientific and Industrial Research-Central Salt and Marine Chemicals Research Institute), Bhavnagar, India. The biomass was then cleaned and shade-dried.

2.2. Mineral Content Analysis

The mineral content estimation for this study was performed according to the protocol of Santoso et al. [22]. A 100 mg of dried seaweed sample was weighed and 5 mL of concentrated nitric acid was added to it. This mixture was allowed to stand overnight. A total of 1 mL concentrated perchloric acid and 100 μL of sulfuric acid were then added to the sample. This was then followed by heating until the emission of white smoke was not observed. A total of 5 mL 2% HCl was used to dissolve the digested sample. The sample was further filtered using a 0.22 μm membrane filter. Inductively coupled plasma atomic emission spectroscopy (ICP-AES) (Perkin-Elmer, Optima 2000, Waltham, MA, USA) was then used for performing the mineralogical analysis of the micro- as well as macro-elements present in *J. rubens*.

2.3. Extraction and Identification of Metabolites

50 mg macro algal sample (DW) was macerated using liquid N_2, followed by the addition of chilled aqueous methanol (70%, v/v) for the determination of secondary metabolites, as per De Vos et al. [23]. The mixture was then vortexed and incubated in an ultrasonic water bath (MRC, Holon, Israel) for 1 h at frequency 40 kHz (25 °C). Thereafter, it was subjected to centrifugation for 10 min at $16,000 \times g$ at 25 °C. The supernatant was collected and filtered using a 0.25 μm membrane. The secondary metabolite analysis was done using Liquid Chromatography coupled with Time of Flight Mass Spectrometry (Micromass, Waters, Milford, MA, USA). The identification of these metabolites was performed by comparing the LC-TOF MS/MS peaks to the free METLIN database [24]. The source and desolvation temperatures were adjusted to 110 and 200 °C, respectively. A total of 2.5 kV was applied to the electrospray capillary, with the cone voltage kept as 25 V, and with nitrogen used as the collision gas. The filtered sample was then directly injected using a syringe pump to the ESI-MS at a flow rate of 50 μL min^{-1}. The extracted metabolites were examined in positive-ion ESI/MS-MS mode. The scanning range was 0–1000 m/z, with an acquisition rate of 0.25 s, and an inter-scan delay of 0.1 s. For the peak integration, the background of each spectrum was subtracted, the data was smoothed and centered, and peaks were integrated using Mass Lynx software version 4.0 (Micromass, Waters, Milford, MA, USA).

2.4. Estimation of Non-Enzymatic Antioxidant Potential

The determination of the antioxidant potential of the *J. rubens* extract was measured as per Gupta et al. [25] with slight modification. 10 g of the dry sample was milled and then extracted twice with 50 mL of MeOH at room temperature for 48 h. The mixture was then centrifuged at 10,000× *g* for 15 min. This extraction step was repeated twice. The supernatants were then collected, combined, filtered, and evaporated to dryness using a rotary evaporator (Buchi Rota vapor R-200, Flawil, Switzerland) under vacuum at 37 °C (150–100 mbar) to give the methanolic extract (dark green mass). The extract was then used to define the total flavonoid content, total phenolic content as well as the antioxidant potential. All the tests were performed in triplicates and the values have been expressed as mean ± SD.

2.4.1. Total Phenolic Content

The total phenolic content in this study was determined following the protocol as per Lim et al. [26]. A total of 2.9 mL of Milli Q water and 0.5 mL of Folin-Ciocalteu's reagent were mixed with the seaweed extract. After 10 min of incubation, 1.5 mL of 20% sodium carbonate solution was added. All the contents were mixed carefully and allowed to stand for 1 h under dark conditions at room temperature. The absorbance was measured at 725 nm. The total phenolic content was calculated based on a standard curve of phloroglucinol.

2.4.2. Total Flavonoid Content (TFC)

The determination of total flavonoid content was conducted according to Zhishen et al. [27]. Briefly, the methanolic extract of *J. rubens* was added to 5% (v/v) $NaNO_2$ (0.3 mL) and incubated for 5 min at room temperature. Thereafter, 2 mL NaOH (1 M) and 0.3 mL $AlCl_3$ (10%, v/v) were added. The absorbance was recorded at 510 nm. TFC was expressed as mg quercetin equivalents (QE)/g extracts (DW).

2.5. DPPH Radical Scavenging Assay

A simple yet sensitive procedure used for the antioxidant studies in natural product chemistry is the DPPH radical scavenging assay. The preparation of the working solution was done by diluting the DPPH stock solution (0.024%, w/v in methanol) until an absorbance of (0.98 ± 0.02) was achieved at 517 nm. The assay was performed using a 96-well plate with five different concentrations of the seaweed extract, ranging from 1250–2500 μg/mL. The extracts were mixed with the DPPH working solution and incubated in dark, for facilitating the color change from purple to yellow upon absorption of hydrogen from the extracts. All the concentrations were tested in triplicates. Ascorbic acid was used as the standard. The absorbance was noted at 517 nm and the scavenging potential of the extracts was calculated using the following equation:

$$\text{Scavenging capacity (\%)} = \frac{OD_{517} \text{ of Control} - OD_{517} \text{ of Extract} \times 100}{OD_{517} \text{ of Control}}$$

2.6. Ferric Ion Reducing Antioxidant Power (FRAP) Assay

For this study, stock solutions of 10 mM Tripyridyltriazine (TPTZ) solution in 40 mM hydrochloric acid, 300 mM acetate buffer with pH 3.6 (16 mL glacial acetic acid + 3.1 g sodium acetate trihydrate), and 20 mM ferric chloride hexahydrate solution were prepared. A fresh working solution was prepared each time, by mixing 25 mL acetate buffer, 2.5 mL $FeCl_3 \cdot 6H_2O$, and 2.5 mL TPTZ solution. Before use, this mixture was heated. *J. rubens* extracts with different concentrations (1250–2500 μg/mL) were allowed to react with the FRAP solution in the dark, in a 96-well titer plate for 30 min. The formation of colored products, ferrous tripyridyltriazine complex was observed and the readings were noted at an absorbance of 593 nm. The results have were expressed in terms of mg Ascorbic Acid Equivalents/g seaweed (DW).

2.7. ABTS Scavenging Activity

The scavenging activity of 2,2′-azino-bis(3-ethylbenz-thiazoline-6-sulfonic acid (ABTS) was determined by using different concentrations of seaweed extracts (1250–2500 µg·mL^{-1}). A working solution was prepared by mixing 50% of Reagent 2 (2.6 mM Potassium persulfate) + 50% of Reagent 1 (7.4 mM ABTS). In order to facilitate the formation of 2,2′-azino-bis(3-ethylbenzothiazoline-6-sulfonic acid) ABTS cations, this mixture was incubated in the dark for 12 h. This assay was based on the capacity of the extract (with different concentrations) to scavenge these ABTS cations. An aliquot of 1 mL from this prepared solution was added to 59 mL methanol and the optical density (OD) value was set as 1.1 at an absorbance of 734 nm. The test samples were then made to react with this solution in a 96-well micro titer plate and further incubated for 2 h in dark. The absorbance was then noted down at 734 nm against the blank, using trolox as a standard, and the scavenging activity was determined. The inhibition (%) was calculated as per the formula:

$$\text{Inhibition } (\%) = \frac{\text{OD}_{734} \text{ of Control} - \text{OD}_{734} \text{ of Extract} \times 100}{\text{OD}_{734} \text{ of Control}}$$

2.8. Total Antioxidant Assay

The total antioxidant activity of the seaweed extracts with different concentration (1250–2500 µg·mL^{-1}) was assessed as per Prieto et al. [28] with certain modifications. The protocol was standardized for a 96-well plate method, for ease. The total antioxidant reagent was prepared by mixing 0.6 M H$_2$SO$_4$ + 28 mM sodium phosphate + 4 mM ammonium molybdate. The samples were mixed with the prepared reagent, and kept at 95 °C for 90 min. The absorbance was recorded at 695 nm. Trolox was used as a standard. The standard curve of the trolox solution was prepared following a similar procedure.

3. Results

3.1. Proximate Composition

The in-depth mineral composition analysis of *J. rubens* was done in order to decipher its cosmeceutical potential. The investigated algal species (in triplicate) contained a good quantity of macro-elements (Na, K, Ca and Mg), as well as microelements (Fe, Mn, Zn, Cu, Mo, and Ni) on a dry weight basis (Table 1). The total quantities of the macro minerals (Na + K + Ca + Mg) obtained were 18,860.74 ± 1.60 mg/100 g DW, while that for micro minerals (Fe + Cu + Zn + Mn + Ni) was 88.15 ± 0.89 mg/100 g DW. Among the macro minerals, Ca content was found to be present in the highest amount (14790.33 ± 1.46 mg/100 g DW), while K content was 627.33 ± 1.52 mg/100 g DW, the lowest whereas among the micro minerals, Fe content was the highest (84.93 ± 0.89 mg/100 g DW) and Cu content was the lowest (0.63 ± 0.05 mg/100 g DW). Other macro minerals such as Na and Mg contents were 1215 ± 1.00 mg/100 g DW and 2228.08 ± 2.09 mg/100 g DW respectively. Micro minerals such as Zn and Mn were present in trace quantities (1.35 ± 0.12 mg/100 g DW and 1.23 ± 0.15 mg/100 g DW) respectively, while Ni showed a value below detectable levels.

3.2. Total Phenol Content

The presence of natural antioxidants is not simply restricted to terrestrial sources. Seaweeds have also proven to be reliable and rich sources of natural antioxidant compounds [29]. The radical scavenging properties of the macro algal species have been reported to be associated with the phenolic compounds present within them [30]. Phenolics possess unique redox properties that contribute in adsorbing as well as neutralizing free radicals, decompose peroxides, and quench singlet and/or triplet oxygen. The antioxidant activity of phenolics primarily owe their properties to such features. Figure 1a represents the total phenol content of the methanolic extract (highest concentration) of *J. rubens* as determined by using the Folin-Ciocalteu reagent and expressed as mg phloroglucinol equivalents

(PGE) per g of seaweed extract, on a dry weight basis. A high total phenolic content was observed to be present in the methanolic extract of *J. rubens*, although this was not higher than the commercial control (Phloroglucinol) used at the same concentration.

Table 1. Mineral composition study of *J. rubens*.

Mineral Composition	
Macro Minerals (mg $(100\ g)^{-1}$ DW)	
Na	1215 ± 1.00
K	627.33 ± 1.52
Ca	$14,790.33 \pm 1.46$
Mg	2228.08 ± 2.09
Na + K + Ca + Mg	$18,860.74 \pm 1.60$
Micro Minerals (mg $(100\ g)^{-1}$ DW)	
Fe	84.93 ± 0.89
Cu	0.63 ± 0.05
Zn	1.35 ± 0.12
Mn	1.23 ± 0.15
Ni	Nd
Fe + Cu + Zn + Mn + Ni	88.15 ± 0.89
Total (Macro + Micro minerals)	$18,948.89 \pm 1.06$ (mg $(100\ g)^{-1}$ DW)

3.3. Total Flavonoid Content

Flavonoids comprise the principal class of polyphenols. They can scavenge practically all known ROS, depending on their structure. They possess the capability to not only scavenge free radicals, but also inhibit the enzymes responsible for free radical production, and also chelate metal ions like iron and copper. The antioxidative assets of the flavonoids may be thus be credited to such mechanisms. Figure 1b represents the total flavonoid content of *J. rubens*. Quercetin was used as a standard, and the total flavonoid content was expressed as mg quercetin equivalents (QE) per g of the seaweed extract on dry weight basis. The total flavonoid content of the methanolic extract of *J. rubens* was thus, found to be lower as compared to the standard when used at a similar concentration (Figure 1b).

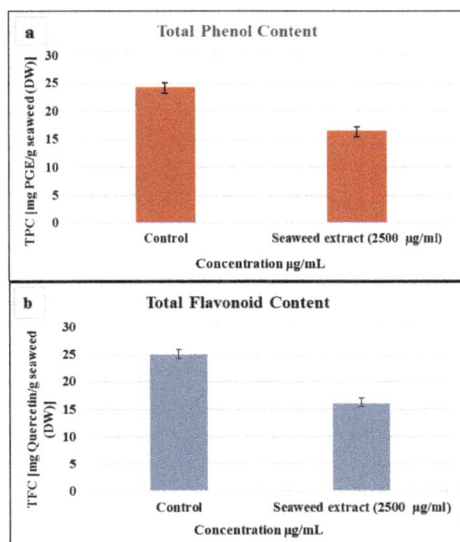

Figure 1. (**a**) Total phenol content and (**b**) Total flavonoid content of *J. rubens* extract at highest concentration. Value represents the mean + SE.

3.4. DPPH Radical Scavenging Assay

DPPH (1,1-diphenyl-2-picrylhydrazyl) behaves as a free radical donor, and is widely used to assess the radical scavenging potential of antioxidants from natural sources. It possesses a nitrogen free radical and is readily quenched by a free radical scavenger. The DPPH assay is exclusively used as a non-enzymatic antioxidant protocol in determining the radical-scavenging capacity of novel antioxidants of natural origin in organic environments that function as a proton radical scavengers or hydrogen donors. Radical quenching, which can be achieved either by single electron transfer or by hydrogen atoms, result in the neutralization of the DPPH radical. Upon reduction, color changes from purple to yellow with the absorption of hydrogen moiety from the antioxidant. This reaction is stoichiometric in nature, and the antioxidant effect can be easily measured by a decrease in ultra-violet (UV) absorption at 517 nm. The DPPH radical-scavenging capacity of the tropical seaweed *J. rubens* selected for the current study has been shown in Figure 2a. The scavenging activity of the sample is indicated by the degree of discoloration. The scavenging potential increased concomitantly with an increase in the extract concentration. The study showed scavenging activity >60% at the highest concentration of 2500 µg/mL. However, the value obtained was less when compared to the standard ascorbic acid value (90%) used at a similar concentration.

3.5. Ferric Ion Reducing Antioxidant Power (FRAP) Assay

The principal of the FRAP assay is based on the reduction of ferric-tripyridyltriazine (Fe^{3+}-TPTZ) complex to ferrous tripyridyltriazine (Fe^{2+}-TPTZ) by the antioxidants of the sample at low pH. Fe^{2+}-TPTZ—the final end product, shows a blue color with an absorption maximum at 593 nm. The change in the absorbance is related to the antioxidant capacity of the seaweed extract. The reducing potential of *J. rubens* extract has been shown in Figure 2b. The reduction potential has been expressed as (µg/mL AAE/mL). The study showed a reduction potential value >50% at the highest concentration of 2500 µg/mL for the extract. However, the value obtained was less when compared to the standard ascorbic acid value (>60%) at the same concentration.

3.6. ABTS Scavenging Activity

The principal underlying this assay is that the ABTS is converted to its radical cation on the addition of potassium persulfate. The radical cation so formed is blue in color and absorbs light at 734 nm. This ABTS radical cation is reactive towards a majority of antioxidants which include thiols, phenolics and ascorbic acid. The blue colored ABTS radical cation is converted back to its colorless neutral form during the course of the reaction. The antioxidants in the sample reduce ABTS, preventing the color formation to a level that is proportional to their concentrations. The ABTS scavenging potential of the *J. rubens* extract is shown in Figure 2c. The results obtained at the highest concentration (2500 µg/mL) were quite comparable to that obtained for the standard (>90%) at similar concentration.

3.7. Total Antioxidant Assay

The total antioxidant capacity (TAC) was performed using the phosphomolybdenum method. The principle behind this protocol is the formation of Mo (V) by the reduction of Mo (VI) using the test extract/s, and consequent production of the green phosphate/Mo (V) complex at an acidic pH. It evaluates both fat-soluble as well as water-soluble antioxidants (total antioxidant capacity). The phosphomolybdenum method is a simple, cheap, and a good alternative to other laborious methods used for assessing the total antioxidant capacity. This assay was carried out in order to gain a broader understanding of the total antioxidant capacity exhibited by *J. rubens*, rather than determining the antioxidant potential of individual constituents. Moreover, it gives a clear understanding of the changes in the antioxidant activity in relation to the oxidative stress, and is widely used in a variety of complex mixtures used in pharmaceutical and cosmetic preparations. The results for the TAC have been expressed as scavenging %, as shown in Figure 2d. The total antioxidant capacity of the extract (>65%) was found to be lower as compared to the standard (>75%) at the same concentration (2500 µg/mL).

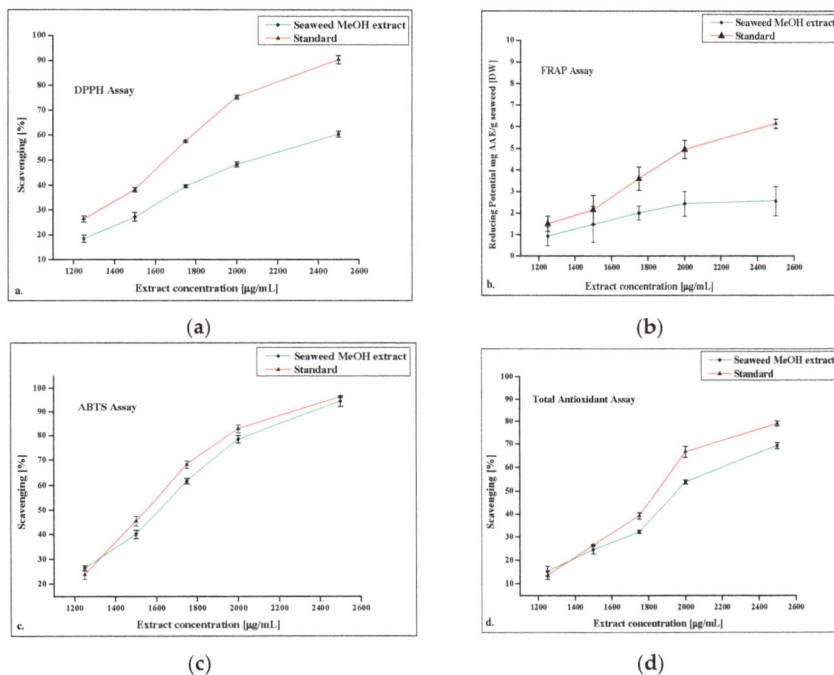

(a)

(b)

(c)

(d)

Figure 2. Scavenging and reducing activities of *J. rubens* extract. (**a**) DPPH (1,1-diphenyl-2-picrylhydrazyl) scavenging assay (**b**) FRAP (Ferric Ion Reducing Antioxidant Power) assay (**c**) ABTS (2,2′azino-bis (3 ethylbenzothiazoline-6-sulfonic acid) assay and (**d**) Total antioxidant assay. Value represents the mean ± SE.

3.8. Metabolite Profiling

Using liquid chromatography-TOF-MS, around 23 different probable metabolites were encountered in the ESI positive mode (Figure 3). These secondary metabolites comprised of alkaloids, phenyl propanoid, furanoid lignans, flavonoids, tannin, terpene lactone, glycosyl and glycoside compounds, and malleolides and analogues, etc. A phytochemical piperidine alkaloid γ-coniceine (m/z 108.12), which is a precursor of several hemlock alkaloids, was detected. It is used in several formulations that are used to treat breast cancer [31]. Another alkaloid, berberrubin (m/z 305.1), bearing strong antitumor activity was also observed in the seaweed. A potential aromatic compound belonging to the class of terpene glycosides named tsangane L3 glucoside (prenol lipid containing a carbohydrate moiety glycosidically bound to a terpene backbone) with m/z 357.23, was detected. The industrial application of this compound is as a surfactant, as well as an emulsifier. An aromatic coumaric acid derivative named trichocarposide (m/z 415.15), was observed. Trichocarposide has been found to be one of the active ingredients in the methanolic extract of *Salix martiana* Leyb that exhibits strong DPPH antioxidant activity [32]. A diterpene-based alkaloid, icaceine (m/z 358.24), known to exhibit anticonvulsant properties, was detected. Two aromatic heteropolycyclic compounds viz; 8-Acetoxy-4′-methoxypinoresinol (a furanoid lignan, and a constituent of *Olea europaea* (olive), which is widely used in cosmetics and pharmaceutical industries), sesartemin (a phenylpropanoid and an inhibitor of cytochrome P450-linked oxygenase) [33] and a natural lignan compound cubebinone (an active constituent of *Piper cubeba* possessing anti-inflammatory as well as antiseptic properties), all with an (m/z 413.16), were detected during the study. Four flavonoids, viz, agecorynin C, medicocarpin, 4′-*O*-methylglucoliquiritigenin, and 5,6,7,8,3′,4′,5′-heptamethoxyflavone, a crude drug, viz, zizybeoside I, and an *O*-glycosyl compound named benzyl gentiobioside (all with an m/z 415.15) were detected. 6-*O*-Methylarmillaridin (m/z 445.18), belonging to the family of melleolides and analogues, was detected to be present in the current study.

An eremophilane-type sesquiterpene lactone, namely eremopetasitenin B2 (m/z 447.18)—an important biogenetic intermediate, and eremosulphoxinolide A—a terpene lactone (m/z 447.18), along with a terpenoid, viz, agnuside (m/z 449.15), an active constituent playing a role in the estrogenic activity for the potential treatment of premenstrual syndrome in women [34], were found to be present. Additionally, two anthocyanins malonylshisonin/4‴-demalonylsalvianin (m/z 825.19) belonging to the class of flavonoids, were also detected which possess antioxidant and anticancer (chemo preventive) attributes. Maclurin 3-C-(2″,3″,6″-trigalloylglucoside) (m/z 863.15) belonging to the family of hydrolysable tannins, and the class of chemical entities known as phenolic glycosides, were identified. Theaflavonin (m/z 919.18), belonging to the class of organic compounds known as complex tannins, was detected. An aromatic coumaric acid ester derivative, simmondsin 2′-ferulate (m/z 534.21), was also observed to be present.

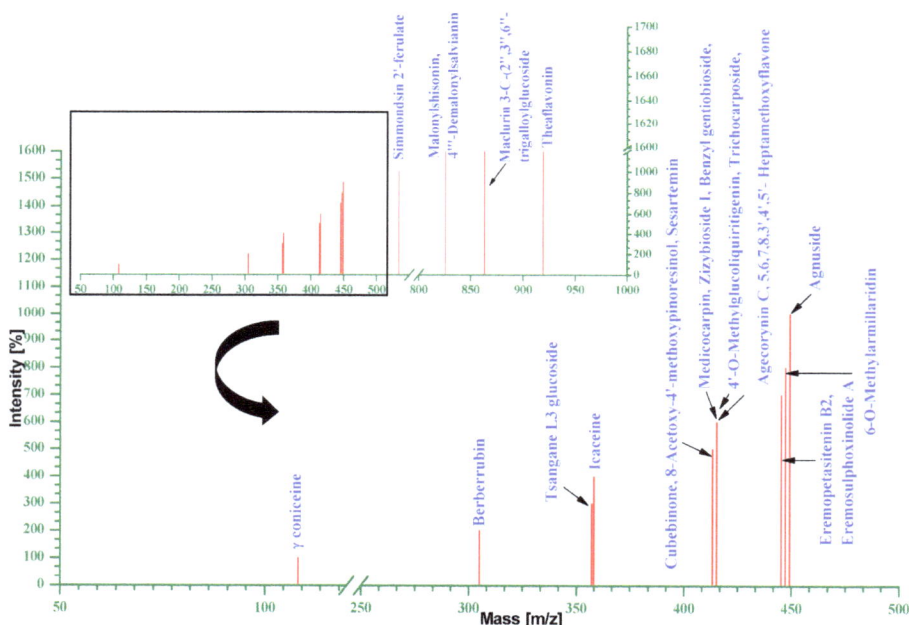

Figure 3. Metabolites detected from *J. rubens* extract using Liquid Chromatography–Mass Spectrometry analysis.

4. Discussion

Jania rubens is a versatile red marine macro algae possessing dichotomous ramifications and extraordinary properties of biomineralization. It is described as an alga that is heavily impregnated with biochemically precipitated $CaCO_3$ in the form of calcite, through the phenomenon of biomineralization. This attribute makes the algae very tough and resistant to corrosion, making the coralline algae one of the most important structural elements of coral reefs. It has an extraordinary capability to bind trace minerals and elements found in seawater, which render it with the invaluable cosmetic virtues that it possesses. The occurrence of coralline algae in several types of habitats defines its reliability as a biological indicator of relative levels of phosphate pollution in seawater caused by human activity [35]. The identification of untargeted metabolites along with the study on the mineral composition, as well as the ROS scavenging capacity of this seaweed, revealed its potential to be used as a key ingredient in the cosmeceutical formulations, and provide useful insights into the metabolite constituents of this macro alga.

The health as well as the appeal of the skin depends on nutrition [36]. This species is a noble source of both macro- as well as micro-minerals. Macro-minerals (Ca, Na, Mg and K) were found to be present in significant amounts. Sodium (Na), plays a crucial role in the fluid maintenance, muscle contraction, enzyme operation, and osmo-regulation of the human body. Additionally, it also plays a critical role in skincare also, due to its anti-aging properties. This anti-aging feature can be attributed to its ability to fight the free radicals that accelerate the aging process. It is therefore incorporated as an active ingredient in a range of skincare products. It is also widely used in cleansers and moisturizers. Products for sensitive skin care also contain sodium, as it acts as a mild wetting agent. Thus, sodium in combination with other elements or alone, offers a number of benefits for the skin. Potassium (K), an essential macro mineral, keeps the skin hydrated and moisturized. Additionally, potassium prevents the skin from looking dull and cracked, as it supports the growth of new skin cells. Many skin care and cosmetic products use KOH (potassium hydroxide) as one of their composites. It maintains a balance in the pH level of the skin. Calcium (Ca) was found to be present in the highest amount as a macro mineral in this species. Ca has been proven to play an important role in skin homeostasis (self-replenishing process) and barrier function repair [37]. Due to this, the skin can shed and renew itself and maintain an appropriate lipid level. Calcium skin benefits include anti-aging properties, enabling a better resistance to irregularities like premature aging and fine wrinkling, through its important role in antioxidant production. Magnesium (Mg) possesses the capacity to cleanse the skin and detoxify the epidermis. It is quite effective in treating the areas of the skin that are prone to allergic reactions. Magnesium is very effective in reducing wrinkles and fine lines. It helps to combat acne or breakouts on the skin. Magnesium thus acts as a natural cellular protectant, fosters the restoration of cellular magnesium levels, facilitates effective and safe detoxification, provides relief from pains, spasms, aches and in turn, encourages healthy skin tissue growth.

Micro minerals also play an equally significant role in maintaining the youthfulness of skin. Iron (Fe) is a potential therapeutic target in the skin [38]. Copper (Cu) too plays an important role in skin care. It not only aids in the production of melanin which is responsible for the color of the skin and hair, but also helps in collagen production and skin regeneration, and increases the effect of antioxidants. Moreover, copper peptides (copper bound peptides) and copper gluconates (copper bound gluconic acid) are a promising treatment in skincare. Zinc (Zn) in the form of divalent zinc ions, provides an antioxidant photoprotector for skin. The benefits of either oral or topical zinc in the treatment of acne, possibly through anti-inflammatory effects, has been studied previously [39]. "Prolidase", an enzyme that is necessary for collagen production—an essential structural component of the skin—requires manganese (Mn) as a co-factor. Manganese is thus an important mineral for everyday skin health, as it plays a specific role in collagen production. In addition to this, manganese aids in the protection of skin against oxygen-related damages, as well as against damages caused due to exposure to UV light by functioning as an antioxidant. *J. rubens* possesses all of these vital macro- as well as micro-minerals crucial for providing protection against pre-mature aging and maintaining the youthfulness of the skin.

One of the most fundamental entities that are known to be involved in human skin aging process are the ROS. The factors that are responsible for ROS generation within the skin include both, the intrinsic sources such as endogenous oxidative metabolism as well as the extrinsic sources such as UV radiation. The role of secondary metabolites is thus important for combating against ROS, as they are considered to be important radical scavengers and efficient antioxidants that possess biological activities [40]. Phytochemicals (secondary metabolites), including flavonoids, phenolic acids, and polyphenols, are potent antioxidants and are an essential part of human well-being. Antioxidants are important as a part of an anti-ageing skin support program because antioxidants help protect the skin from the toxic effects of free radicals that would otherwise impair and destroy healthy skin cells [41]. Scavenging properties and antioxidant potential depend on the content of both polyphenolics and flavonoids [40]. *J. rubens* possessed a good amount of both phenolic as well as flavonoid contents, and therefore showed a high antioxidant potential. The antioxidant activity was found to be proportional to the concentration of polyphenols and flavonoids. Five different methanolic concentrations were used to check the DPPH radical scavenging potential. The radical scavenging potential was lower compared to the standard ascorbic acid but was >60%. This result suggests the presence of free radical inhibitors acting as potential primary antioxidants. Therefore, *J. rubens* extract was found to contain metabolites that could function as hydrogen donors, thereby neutralizing the DPPH molecules. Moreover, the presence of metabolites like trichocarposide, that exhibit strong DPPH antioxidant activity, further supported the study. In the reducing power assay, the antioxidants donate an electron to stabilize the radicals and also break the free radical chain reaction. The ability of the different concentrations of the *J. rubens* extract to exhibit the reducing power in this investigation may be related to the presence of antioxidant phytochemicals. Reducing potential was found to be lower as compared to the commercial standard. Some of the cosmeceutical effects could be attributed to these compounds. The ABTS radical scavenging ability of the extract also exhibited a dose-dependent response. As the concentration of the methanolic extracts increased, ABTS radical scavenging as well as the total antioxidant activity also increased. In fact, the ABTS radical scavenging potential was found to be quite comparable to the standard value, while the latter was lower as compared to the standard.

Furthermore, about 23 different putative metabolites were identified in the present study using LC-TOF-MS/MS (Figure 3 and Supplementary Table S1). This study offers the first comprehensive scientific report on the untargeted metabolomics of *J. rubens*, highlighting its importance as an active cosmeceutical ingredient. Anthocyanins [42,43] provide protection against potentially effective agents, in order to avert the signs of skin aging. They have proven actions for giving protection to the skin from external injuries caused by UV radiation [44]. Additionally, anthocyanins not only possess strong antioxidant/anti-inflammatory activities, but also inhibit lipid peroxidation and inflammatory mediators; cyclooxygenase (COX)-1 and -2 [45]. They are also used as cosmetic colorants in many products [46]. The present study unveils the presence of two such anthocyanins, malonylshisonin and 4′′′-demalonylsalvianin, which may prove to be potential flavonoids for cosmetic industries. A complex oxidation product belonging to the natural class of flavonoid theaflavins, named theaflavonin, was detected. Theaflavins are potent antioxidant polyphenols already used in skin care product preparation, in order to impart effects such as whitening and the removal of freckles. [47]. Moreover, theaflavin in the skin care products shows excellent color stability. Therefore, they can be easily applied to other cosmetic preparations as well. Flavonoids have been found to possess antioxidant, anti-allergic, anti-viral, anti-aging, anti-carcinogenic, and anti-inflammatory properties. Moreover, they are also used in the treatment of skin aging, as they apparently contribute in the improvement of skin elasticity, skin hydration, regulation of oil gland secretion, and collagen content [48]. A set of flavonoids, namely agecorynin C (KEGG C14942), medicocarpin (KEGG C16223), and 5,6,7,8,3′,4′,5′-heptamethoxyflavone (KEGG C14953), have been detected in this study. Benzyl gentiobioside, belonging to the class of *O*-glycosyl compounds, was also detected.

Cosmetic formulations also possess natural as well as synthetic additives that confer aroma, which in turn masks unpleasant chemical odors. This study also revealed the presence of many natural aromatic compounds. One of them was trichocarposide—an aromatic coumaric ester derivative that is an essential constituent of *Populus balsamifera* (balsam poplar), which is highly sought after for its essential oil in various aroma therapies, was detected from *J. rubens* extract. It is known to nourish the skin and relax the mind. Additionally, 4′-*O*-Methylglucoliquiritigenin—a compound belonging to flavonoid-7-*O*-glycosides, was also detected. It is one of the constituent of the roots of *Glycyrrhiza uralensis* (Chinese licorice). The pharmacological significance of licorice includes its antimicrobial [49], antiviral [50], and antitumor [51], as well as its anti-inflammatory [52] properties. Zizybeoside I (KEGG C17564), detected during the present study, is usually found to be present in *Zizyphus jujuba* (Chinese date), and belongs to the family of Dihexoses (disaccharides containing two hexose carbohydrates). *Zizyphus jujuba* is commonly used for face nourishment and beautification, for battling oxidation and aging, as an antitumor agent, for resisting fatigue, and also for the relaxation of mind [53–55]. In fact, Chinese jujube is also commonly employed in Chinese beverages and as a food additive [56].

Eremopetasitenin B2 and eremosulphoxinolide A, isolated from the fresh rhizomes of *Petasites japonicus*, which has been used for its therapeutic effect on allergy and asthma in Korea and European countries, was also detected [57,58]. Simmondsin 2-ferulate, a glucoside with proven insecticidal, antifeedant, and antifungal activities [59] was noted to be present. Simmondsin 2′-ferulate is usually obtained from *Simmondsia chinensis* (jojoba). Lubricant, pharmaceutical and cosmetic industries hold a good market for Jojoba oil [60]. Jojoba seeds have also been shown to possess anti-inflammatory activity [61].

Terpenes comprise one of the major secondary metabolites, with diverse categories including monoterpenoids, diterpenoids, triterpenoids, and sesquiterpenoids. Terpenes have been found to possess anti-inflammatory, antitumor, antibacterial, antioxidant, and hepatoprotective activities from various pharmacological studies [62]. Tsangane L3 glucoside (prenol lipid containing a carbohydrate moiety glycosidically bound to a terpene backbone)—a potential aromatic compound belonging to the class of terpene glycosides, was detected in the present study. Another diterpene-based alkaloid, named icaceine (m/z 358.24), known to prevent or reduce the severity of epileptic seizures and agnuside, an iridoid glycoside possessing hepatoprotective properties [63], anti-inflammatory activity [64] antioxidant [65], and analgesic effects [66], were found to be present in the methanolic extract of *J. rubens*. Additionally, two alkaloids viz; γ-coniceine (KEGG C10138) which acts as a local analgesic [67], and berberrubin (HMDB30266), a protoberberine, known for exhibiting antitumor properties, were detected during the study [68].

Lignans act as both antioxidants and phytoestrogens [69]. Other salient traits, viz, anti-cancerous, anti-inflammatory, insecticidal, anti-hypertensive, hypocholesterolemic, anti-asthma activities, and hepatoprotective biological activities of lignans have also been previously reported [70]. Lignans derived from flax have not only been described as being useful in preventing osteoporosis and certain cancers, but also as anti-viral agents and fungicides. The role of lignans in plant defense was suggested by their pronounced antimicrobial, antifungal, antiviral, and antioxidant properties [71–74]. The glycosylated lignans obtained from flax seeds are known to inhibit the production of melanin, thereby playing a role in bleaching the skin [75]. They also act as agents for increasing the catalase and fibroblast activity, for inhibiting UV-induced wrinkles, for the synthesis of hyaluronic acid, and for strengthening the elasticity of the skin. Further, the prevention of formation of wrinkles, thereby assisting with maintaining the firmness of the skin by providing relief through scavenging OH radicals, has also been proven outcomes for glycosylated lignans derived from sesame. These are also well known as antioxidants, and may be used in moisturizing and protective creams [76]. The retinol-based anti-ageing compositions contain lignan known as nordihydroguaiaretic acid, which is an antioxidant. A furanoid lignan 8-acetoxy-4′-methoxypinoresinol (HMDB33277), and two natural lignin compounds, sesartemin (KEGG C10884) and cubebinone, (HMDB33259) were observed to be present in the methanolic extract of *J. rubens*. Certain melleolide analogs have shown to exhibit anti-microbial as well as antifungal activities [77,78]. 6-*O*-methylarmillaridin, a melleolide sesquiterpene was one of the bioactive metabolites encountered during the present study.

A number of brown seaweeds viz; *Fucus vesiculosus* (Fucoidan) [79], *Pelvetia wrightii* [80], *Laminaria digitata* (Minerals, proteins and carbohydrates) [81], are frequently used in slimming and anticellulite formulations. Macro algal constituents like flavonoids, phlorotannins, quercetin (flavonoid belonging to the class of flavonols) etc. act as lipolytic agents. *Sargassum polycystum* extracts (ethanolic, ethyl acetate and hexane) have been studied for their skin whitening properties [82]. The biochemical composition of these active fractions have revealed the presence of terpenoids, flavonoids, phenols, saponins, tannins etc. Moreover, antioxidants favor the skin health by reducing hyperpigmentation. It has been reported that macro algae have emerged with mechanisms to respond against the hazardous effects of UV-A and UV-B by producing phenolic compounds, UV-absorbing mycosporine-like amino acid (MAAs), or carotenoids [79]. Such MAAs act as a UV shield, and the antioxidant constituents present in the red algae (Rhodophytes) act as effective photoprotectors. Similarly, in the case of microalgae, an extract from *Chlorella vulgaris* has been reported to induce collagen synthesis in skin, thereby preventing wrinkle formation [83]. A protein-rich extract from *Arthrospira* has been reported to exert tightening of the skin, thereby protecting the early signs of skin aging [83]. The current study too has revealed the presence of many such bioactive metabolites in *J. rubens*, which may open up new avenues for the advancement in cosmetic compositions.

5. Conclusions

The study reveals that the red marine macro alga, *J. rubens*, is a rich source of essential macro- as well as micro-minerals, natural antioxidants, and bioactive metabolites with cosmeceutical potential. These features collectively make this coralline alga a promising candidate for its inclusion as an active ingredient in modern day cosmeceutical as well as pharmaceutical formulations/products utilized for skin conditioning, skin polishing, anti-ageing, and skin whitening. *J. rubens* can therefore be considered as a potential natural candidate for improving the quality of modern day cosmetics. However, further studies undertaking purification as well as clinical trials of the purified components, are required for its inclusion as an integral ingredient. The present study provides the base for future perspectives of considering *J. rubens* in the design of algal biorefineries and other eco-designs, to obtain large amounts of value-added products.

Supplementary Materials: The following are available online at www.mdpi.com/2079-9284/4/4/45/s1, Figure S1: Metabolite Profile of *J. rubens* extract using LCMS (ESI positive mode), Table S1: Putative metabolites identified in *J. rubens* extract and their possible function/application/role/importance.

Acknowledgments: Dhara Dixit (D.D.) sincerely thanks the Department of Science & Technology (DST), New Delhi, for awarding the Women Scientist Fellowship and rendering financial support (WOS-A/LS-388/2012) for undertaking this research. Dhara Dixit deeply acknowledges the support rendered by the Director, CSIR-CSMCRI for providing the infrastructure and the analytical facilities to conduct this study. The guidance extended by Avinash Mishra for LC-MS study is highly acknowledged. The field assistance extended for the collection of seaweed sample by Devesh K. Gadhavi (Deputy Director, Kutch Ecological Research Centre—A Division of The Corbett Foundation, Kachchh) is duly acknowledged. The author also thanks M.G. Thakkar (Head) & M.H. Trivedi (Asst. Prof.), Department of Earth & Environmental Science, Krantiguru Shyamji Krishna Verma (K.S.K.V.) Kachchh University for providing the administrative support for carrying out the work smoothly. The fellowship term is over and there is no fund for covering the costs to publish in open access.

Author Contributions: Dhara Dixit conceived and designed the experiments; Dhara Dixit performed the experiments; Dhara Dixit analyzed the data; C.R.K. Reddy contributed reagents/materials/analysis tools; Dhara Dixit and C.R.K. Reddy wrote the paper.

Conflicts of Interest: The authors declare no conflict of interest.

References

1. Thomas, N.V.; Kim, S.K. Beneficial Effects of Marine Algal Compounds in Cosmeceuticals. *Mar. Drugs* **2013**, *11*, 146–164. [CrossRef] [PubMed]
2. Kim, S.K.; Ravichandran, Y.D.; Khan, S.B.; Kim, Y.T. Prospectives of the cosmeceuticals derived from marine organisms. *Biotechnol. Bioprocess Eng.* **2008**, *13*, 511–523. [CrossRef]
3. Wijesinghe, W.A.J.P.; Jeon, Y.J. Biological activities and potential cosmeceutical applications of bioactive components from brown seaweeds: A review. *Phytochem. Rev.* **2011**, *10*, 431–443. [CrossRef]
4. Cabrita, M.; Vale, C.; Rauter, A. Halogenated compounds from marine algae. *Mar. Drugs* **2010**, *8*, 2301–2317. [CrossRef] [PubMed]

5. Manach, C.; Scalbert, A.; Morand, C.; Rémésy, C.; Jiménez, L. Polyphenols: Food sources and bioavailability. *Am. J. Clin. Nutr.* **2004**, *79*, 727–747. [PubMed]

6. Zern, T.L.; Fernandez, M.L. Cardio protective effects of dietary polyphenols. *J. Nutr.* **2005**, *135*, 2291–2294. [PubMed]

7. Dunn, W.B.; Ellis, D.I. Metabolomics: Current analytical platforms and technologies. *Trends Anal. Chem.* **2005**, *24*, 285–294.

8. Naidoo, K.; Machin, M.A.B. Oxidative Stress and Ageing: The Influence of Environmental Pollution, Sunlight and Diet on Skin. *Cosmetics* **2017**, *4*, 4. [CrossRef]

9. Joe, M.J.; Kim, S.N.; Choi, H.Y.; Shin, W.S.; Park, G.M.; Kang, D.W.; Kim, Y.K. The inhibitory effects of eckol and dieckol from Ecklonia stolonifera on the expression of matrix metalloproteinase-1 in human dermal fibroblasts. *Biol. Pharm. Bull.* **2006**, *29*, 1735–1739. [CrossRef] [PubMed]

10. Kim, M.M.; Van Ta, Q.; Mendis, E.; Rajapakse, N.; Jung, W.K.; Byun, H.G.; Jeon, Y.J.; Kim, S.K. Phlorotannins in Ecklonia cava extract inhibit matrix metalloproteinase activity. *Life Sci.* **2006**, *79*, 1436–1443. [CrossRef] [PubMed]

11. Price, R.D.; Berry, M.G.; Navsaria, H.A. Hyaluronic acid: The scientific and clinical evidence. *J. Plast. Reconstr. Aesthet. Surg.* **2007**, *60*, 1110–1119. [CrossRef] [PubMed]

12. Ryu, B.; Qian, Z.J.; Kim, M.M.; Nam, K.W.; Kim, S.K. Anti-photoaging activity and inhibition of matrix metalloproteinase (MMP) by marine red alga, Corallina pilulifera methanol extract. *Radiat. Phys. Chem.* **2009**, *78*, 98–105.

13. Verdy, C.; Branka, J.E.; Mekideche, N. Quantitative assessment of lactate and progerin production in normal human cutaneous cells during normal ageing: Effect of an Alaria esculenta extract. *Int. J. Cosmet. Sci.* **2011**, *33*, 462–466. [CrossRef] [PubMed]

14. Rossano, R.; Ungaro, N.; D'Ambrosio, A.; Liuzzi, G.M.; Riccio, P. Extracting and purifying R-phycoerythrin from Mediterranean red algae Corallina elongata. *J. Biotechnol.* **2003**, *101*, 289–293. [CrossRef]

15. Joly, A.B. *Generos de Algas Marinhas da Costa Atlantica Latino-Americana*; Editora da Universidade de São Paulo: São Paulo, Brazil, 1967; p. 461.

16. Bôas, V.; Bigio, A.; Figueiredo, M.A.D.O. Are anti-fouling effects in coralline algae species specific? *Braz. J. Oceanogr.* **2004**, *52*, 11–18. [CrossRef]

17. Karabay-Yavasoglu, N.U.; Sukatar, A.; Ozdemir, G.; Horzum, Z. Antimicrobial activity of volatile components and various extracts of the red alga *Jania rubens*. *Phytother. Res.* **2007**, *21*, 153–156. [CrossRef] [PubMed]

18. Ktari, L.; Blond, A.; Guyot, M. 16β-Hydroxy-5α-cholestane-3, 6-dione, a novel cytotoxic oxysterol from the red alga *Jania rubens*. *Bioorg. Med. Chem. Lett.* **2000**, *10*, 2563–2565. [CrossRef]

19. El-Din, S.M.M.; El-Ahwany, A.M. El-Ahwany Bioactivity and phytochemical constituents of marine red seaweeds (*Jania rubens*, *Corallina mediterranea* and *Pterocladia capillacea*). *J. Taibah Univ. Sci.* **2016**, *10*, 471–484. [CrossRef]

20. Awad, N.E. Bioactive brominated diterpenes from the marine red alga *Jania rubens* (L.) Lamx. *Phytother Res.* **2004**, *18*, 275–279. [CrossRef] [PubMed]

21. Navarro, D.A.; Stortz, C.A. The system of xylogalactans from the red seaweed *Jania rubens* (Corallinales, Rhodophyta). *Carbohydr. Res.* **2008**, *43*, 2613–2622. [CrossRef] [PubMed]

22. Santoso, J.; Gunji, S.; Yoshie-Stark, Y.; Suzuki, T. Mineral contents of Indonesian seaweeds and mineral solubility affected by basic cooking. *Food Sci. Technol. Res.* **2006**, *12*, 59–66. [CrossRef]

23. De Vos, R.C.; Moco, S.; Lommen, A.; Keurentjes, J.J.; Bino, R.J.; Hall, R.D. Untargeted large-scale plant metabolomics using liquid chromatography coupled to mass spectrometry. *Nat. Protoc.* **2007**, *2*, 778–791. [CrossRef] [PubMed]

24. Zhu, Z.J.; Schultz, A.W.; Wang, J.; Johnson, C.H.; Yannone, S.M.; Patti, G.J.; Siuzdak, G. Liquid chromatography quadrupole time-of-flight mass spectrometry characterization of metabolites guided by the METLIN database. *Nat. Protoc.* **2013**, *8*, 451–460. [CrossRef] [PubMed]

25. Kumar, M.; Gupta, V.; Kumari, P.; Reddy, C.R.K.; Jha, B. Assessment of nutrient composition and antioxidant potential of Caulerpaceae seaweeds. *J. Food Compos. Anal.* **2011**, *24*, 270–278. [CrossRef]

26. Lim, S.N.; Cheung, P.C.K.; Ooi, V.E.C.; Ang, P.O. Evaluation of antioxidative activity of extracts from brown seaweed, *Sargassum siliquastrum*. *J. Agric. Food Chem.* **2002**, *50*, 3862–3866. [CrossRef] [PubMed]

27. Jia, Z.; Tang, M.; Wu, J. The determination of flavonoid contents in mulberry and their scavenging effects on superoxide radicals. *Food Chem.* **1999**, *64*, 555–559.

28. Prieto, P.; Pineda, M.; Anguilar, M. Spectrophotometric quantitation of antioxidant capacity through the formation of a Phosphomolybdenum Complex: Specific application to the determination of Vitamin E. *Anal. Biochem.* **1999**, *269*, 337–341. [CrossRef] [PubMed]
29. Patra, J.K.; Lee, S.-W.; Park, J.G.; Baek, K.-H. Antioxidant and Antibacterial Properties of Essential Oil Extracted from an Edible Seaweed *Undaria Pinnatifida*. *J. Food Biochem.* **2017**, *41*, e12278. [CrossRef]
30. Duan, X.J.; Zhang, W.W.; Li, X.M.; Wang, B.G. Evaluation of antioxidant property of extract and fractions obtained from a red alga, *Polysiphonia urceolata*. *Food Chem.* **2006**, *95*, 37–43. [CrossRef]
31. Usman, M.R.M.; Salgar, S.D.; Nagpal, N.; Shaikh, M.Z. *Poisonous Herbal Plants: NA*; Educreation Publishing: New Delhi, India, 2016.
32. Fernandes, C.C.; de Carvalho Cursino, L.M.; de Abreu Pinheiro Novaes, J.; Demetrio, C.A.; Júnior, O.L.P.; Nunez, C.V.; do Amaral, I.L. Salicilatos isolados de folhas e talos de *Salix martiana* Leyb. (Salicaceae). *Quim. Nova* **2009**, *32*, 983–986. [CrossRef]
33. Polya, G. *Biochemical Targets of Plant Bioactive Compounds: A Pharmacological Reference Guide to Sites of Action and Biological Effects*; CRC Press: Boca Raton, FL, USA, 2003; ISBN 0-415-30829-1.
34. Ramakrishna, R.; Bhateria, M.; Singh, R.; Puttrevu, S.K.; Bhatta, R.S. Plasma pharmacokinetics, bioavailability and tissue distribution of agnuside following peroral and intravenous administration in mice using liquid chromatography tandem mass spectrometry. *J. Pharm. Biomed. Anal.* **2016**, *125*, 154–164. [CrossRef] [PubMed]
35. Brown, V.; Ducker, S.C.; Rowan, K.S. The effect of orthophosphate concentration on the growth of articulated coralline algae (Rhodophyta). *Phycologia* **1977**, *16*, 125–131. [CrossRef]
36. Boelsma, E.; Hendriks, H.F.J.; Roza, L. Nutritional skin care: Health effects of micronutrients and fatty acids1–3. *Am. J. Clin. Nutr.* **2001**, *73*, 853–864. [PubMed]
37. Denda, M.; Fuziwara, S.; Inoue, K. Influx of Calcium and Chloride Ions into Epidermal Keratinocytes Regulates Exocytosis of Epidermal Lamellar Bodies and Skin Permeability Barrier Homeostasis. *J. Investig. Dermatol.* **2003**, *121*, 362–367. [CrossRef] [PubMed]
38. Wright, J.A.; Richards, T.; Srai, S.K.S. The role of iron in the skin and cutaneous wound healing. *Front. Pharmacol.* **2014**, *5*, 156. [CrossRef] [PubMed]
39. Solomons, N.W.; Ruz, M.; Duran, C.C. Putative therapeutic roles for zinc. In *Zinc in Human Biology*; Mills, C., Ed.; Springer: London, UK, 1989; pp. 297–321.
40. Shahidi, F.; Ambigaipalan, P. Phenolics and polyphenolics in foods, beverages and spices: Antioxidant activity and health effects—A review. *J. Funct. Foods* **2015**, *18*, 820–897. [CrossRef]
41. Salavkar, S.M.; Tamanekar, R.A.; Athawale, R.B. Antioxidants in skin ageing—Future of dermatology. *Int. J. Green Pharm.* **2011**, *5*, 161–168.
42. Schreckinger, M.E.; Lotton, J.; Lila, M.A.; De Mejia, E.G. Berries from South America: A comprehensive review on chemistry, health potential, and commercialization. *J. Med. Food* **2010**, *13*, 233–246. [CrossRef] [PubMed]
43. Afaq, F.; Zaid, M.A.; Khan, N.; Dreher, M.; Mukhtar, H. Protective effect of pomegranate-derived products on UVB-mediated damage in human reconstituted skin. *Exp. Dermatol.* **2009**, *18*, 553–561. [CrossRef] [PubMed]
44. Afaq, F.; Khan, N.; Syed, D.N.; Mukhtar, H. Oral feeding of pomegranate fruit extract inhibits early biomarkers of UVB radiation-induced carcinogenesis in Skh-1 hairless mouse epidermis. *Photochem. Photobiol.* **2010**, *86*, 1318–1326. [CrossRef] [PubMed]
45. Seeram, N.P.; Cichewicz, R.H.; Chandra, A.; Nair, M.G. Cyclooxygenase inhibitory and antioxidant compounds from crabapple fruits. *J. Agric. Food Chem.* **2003**, *51*, 1948–1951. [CrossRef] [PubMed]
46. Khan, A.S. *Flowering Plants: Structure and Industrial Products*; Wiley: Hoboken, NJ, USA, 2017.
47. Huiling, L.; Xiufang, Y.; Shikang, Z.; Rong, T.; Yunfei, T.; Junhao, K.; Xiaoqiang, C.; Yuping, G. Application of Theaflavin and Skin Care proDuct Containing Theaflavin. Patent No. CN103,083,199 A, 2013.
48. Michalun, M.V.; Dinardo, J.C. *Milady Skin Care and Cosmetic Ingredients Dictionary*, 4th ed.; Cengage Learning: Independence, KY, USA, 2014.
49. Ahn, S.J.; Cho, E.J.; Kim, H.J.; Park, S.N.; Lim, Y.K.; Kook, J.K. The antimicrobial effects of deglycyrrhizinated licorice root extract on *Streptococcus mutans UA159* in both planktonic and biofilm cultures. *Anaerobe* **2012**, *18*, 590–596. [CrossRef] [PubMed]
50. Adianti, M.; Aoki, C.; Komoto, M.; Deng, L.; Shoji, I.; Wahyuni, T.S. Anti-hepatitis C virus compounds obtained from *Glycyrrhiza uralensis* and other Glycyrrhiza species. *Microbiol. Immunol.* **2014**, *58*, 180–187. [CrossRef] [PubMed]

51. Khan, R.; Khan, A.Q.; Lateef, A.; Rehman, M.U.; Tahir, M.; Ali, F. Glycyrrhizic acid suppresses the development of precancerous lesions via regulating the hyper proliferation, inflammation, angiogenesis and apoptosis in the colon of Wistar rats. *PLoS ONE* **2013**, *8*, e56020.

52. Chandrasekaran, C.V.; Deepak, H.B.; Thiyagarajan, P.; Kathiresan, S.; Sangli, G.K.; Deepak, M. Dual inhibitory effect of Glycyrrhiza glabra (GutGard™) on COX and LOX products. *Phytomedicine* **2011**, *18*, 278–284. [CrossRef] [PubMed]

53. Fatemeh, V.; Mohsen, F.M.; Kazem, B. Evaluation of inhibitory effect and apoptosis induction of *Zyzyphus jujube* on tumor cell lines, an in vitro preliminary study. *Cytotechnology* **2008**, *56*, 105–111.

54. Krings, U.; Berger, R.G. Antioxidative activity of some roasted foods. *Food Chem.* **2001**, *72*, 223–229. [CrossRef]

55. Qi, H.M.; Zhang, Q.B.; Zhao, T.T.; Chen, R.; Zhang, H.; Niu, X.Z.; Li, Z. Antioxidative activity of different sulfate content derivatives of polysaccharide extracted from *Ulva pertusa* (Chlorophyta) in vitro. *Int. J. Biol. Macromol.* **2005**, *37*, 195–199. [CrossRef] [PubMed]

56. Wang, S.; Zhang, J.; Zhang, Z.; Gao, W.; Yan, Y.; Li, X.; Liu, C. Identification of Chemical Constituents in the Extract and Rat Serum from Ziziphus Jujuba Mill by HPLC-PDA-ESI-MSn. *Iran. J. Pharm. Res.* **2014**, *13*, 1055–1063. [PubMed]

57. Sok, D.E.; Oh, S.H.; Kim, Y.B.; Kang, H.G.; Kim, M.R. Neuroprotection by extract of *Petasites japonicus* leaves, a traditional vegetable, against oxidative stress in brain of mice challenged with kainic acid. *Eur. J. Nutr.* **2006**, *45*, 61–69. [CrossRef] [PubMed]

58. Choi, O.B. Anti-allergic effects of *Petasites japonicas*. *Korean J. Food Nutr.* **2002**, *5*, 382–385.

59. Abbassy, M.A.; Abdelgaleil, S.A.M.; Belal, A.S.H.; Rasoul, M.A.A.A. Insecticidal, antifeedant and antifungal activities of two glucosides isolated from the seeds of *Simmondsia chinensis*. *Ind. Crop. Prod.* **2007**, *26*, 345–350. [CrossRef]

60. Arya, D.; Khan, S. A Review of *Simmondsia chinensis* (Jojoba) "The Desert Gold": A Multipurpose Oil Seed Crop for Industrial Uses. *J. Pharm. Sci. Res.* **2016**, *8*, 381–389.

61. Malty, R.H.; Naim, A.A.; Khalifa, A.E.; Azizi, M.M.A. Anti-inflammatory effects of jojoba liquid wax in experimental models. *Pharmacol. Res.* **2005**, *51*, 95–105.

62. Yao, J.L.; Fang, S.M.; Liu, R.; Oppong, M.B.; Liu, E.W.; Fan, G.W.; Zhang, H. A Review on the Terpenes from Genus Vitex. *Molecules* **2016**, *21*, 1179. [CrossRef] [PubMed]

63. Atta-ur-Rahman. *Studies in Natural Products Chemistry, Bioactive Natural Products (Part L)*; Elsevier Science: Amsterdam, The Netherlands, 2005; Volume 32, p. 1268.

64. Pandey, A.; Bani, S.; Satti, N.K.; Gupta, B.D.; Suri, K.A. Anti-arthritic activity of agnuside mediated through the down-regulation of inflammatory mediators and cytokines. *Inflamm. Res.* **2012**, *61*, 293–304. [CrossRef] [PubMed]

65. Tiwari, N.; Luqman, S.; Masood, N.; Gupta, M.M. Validated high performance thin layer chromatographic method for simultaneousquantification of major iridoids in *Vitex trifolia* and their antioxidant studies. *J. Pharm. Biomed. Anal.* **2012**, *61*, 207–214. [CrossRef] [PubMed]

66. Okuyama, E.; Fujimori, S.; Yamazaki, M.; Deyama, T. Pharmacologically active components of viticis fructus (*Vitex rotundifolia*). II. The components having analgesic effects. *Chem. Pharm. Bull.* **1998**, *46*, 655–662. [CrossRef] [PubMed]

67. Nelly, A.; Annick, D.D.; Frederic, D. Plants used as remedies antirheumatic and antineuralgic in the traditional medicine of Lebanon. *J. Ethnopharmacol.* **2008**, *120*, 315–334.

68. Jeon, Y.W.; Jung, J.W.; Kang, M.R.; Chung, I.K.; Lee, W.T. NMR Studies on Antitumor Drug Candidates, Berberine and Berberrubine. *Bull. Korean Chem. Soc.* **2002**, *23*, 391–394.

69. Goyal, A.; Sharma, V.; Upadhyay, N.; Gill, S.; Sihag, M. Flax and flaxseed oil: An ancient medicine & modern functional food. *J. Food Sci. Technol.* **2014**, *51*, 1633–1653. [PubMed]

70. Mao, J.; Yu, N.-J.; Yang, Y.; Zhao, Y.-M. Biological activities of dibenzyl butyrolactone lignans, Research advances. *J. Int. Pharm. Res.* **2014**, *41*, 275–281.

71. Chandra, H.; Bishnoi, P.; Yadav, A.; Patni, B.; Mishra, A.P.; Nautiyal, A.R. Antimicrobial Resistance and the Alternative Resources with Special Emphasis on Plant-Based Antimicrobials—A Review. *Plants* **2017**, *6*, 16. [CrossRef] [PubMed]

72. Windayani, N.; Rukayadi, Y.; Hakim, E.H.; Ruslan, K.; Syah, Y.M. Antifungal activity of lignans isolated from *Phyllanthus myrtifoliu Moon* against Fusarium oxysporum. *Curr. Top. Phytochem.* **2014**, *12*, 33–39.

73. Elfahmi, N.V. Phytochemical and Biosynthetic Studies of Lignans, with a Focus on Indonesian Medicinal Plants. Ph.D. Thesis, University Library Groningen, Groningen, The Netherlands, 2006.

74. Lin, X.; Zhou, L.; Li, T.; Brennan, C.; Fu, X.; Liu, R.H. Phenolic content, antioxidant and antiproliferative activities of six varieties of white sesame seeds (*Sesamum indicum* L.). *RSC Adv.* **2017**, *7*, 5751–5758. [CrossRef]

75. Renault, B.; Catroux, P. Cosmetic Use of Lignans. Patent No. EP1,526,832 A1, 4 May 2005.

76. Toshihiko, O.; Keiko, N.; Kyoko, S.; Takemoto, Y.; Kabushiki, K. Protein Composition Derived from Sesame Seed and Use Thereof. U.S. Patent 5,993,795 A, 30 November 1999.

77. Midland, S.L.; Izac, R.R.; Wing, R.M.; Zaki, A.I.; Munnecke, D.E.; Sims, J.J. Melleolide, a new antibiotic from *Armillaria mellea*. *Tetrahedron Lett.* **1982**, *23*, 2515–2518. [CrossRef]

78. Donnelly, D.M.; Abe, F.; Coveney, D.; Fukuda, N.; O'Reilly, J.; Polonsky, J.; Prangé, T. Antibacterial sesquiterpene aryl esters from *Armillaria mellea*. *J. Nat. Prod.* **1985**, *48*, 10–16. [CrossRef] [PubMed]

79. Bedoux, G.; Hardouin, K.; Burlot, A.S.; Bourgougnon, N. Bioactive Components from Seaweeds: Cosmetic Applications and Future Development. *Adv. Bot. Res.* **2014**, *71*, 345–378.

80. Rozkin, M.; Levina, M.N.; Efimov, V.S.; Usov, A.I. The anticoagulant and lipolysis-stimulating activity of polysaccharides from marine brown algae. *Farmakol. Toksikol.* **1991**, *54*, 40–42. [PubMed]

81. Gedouin, A.; Vallee, R.; Morvan, P.Y. Use of Algae Extract to Stimulate the Oxygen Uptake by the Cells Having Lipolytic Effect to Produce ATP Molecules. Patent No. FR 2,879,098 A1, 16 June 2006.

82. Chan, Y.Y.; Kim, K.H.; Cheah, S.H. Inhibitory effects of Sargassum polycystum on tyrosinase activity and melanin formation in B16F10 murine melanoma cells. *J. Ethnopharmacol.* **2011**, *137*, 1183–1188. [CrossRef] [PubMed]

83. Spolaore, P.; Joannis-Cassan, C.; Duran, E.; Isambert, A. Commercial Applications of Microalgae. *J. Biosci. Bioeng.* **2006**, *101*, 201–211. [CrossRef] [PubMed]

cosmetics

MDPI

Article

Castanea sativa Bur: An Undervalued By-Product but a Promising Cosmetic Ingredient

Diana Pinto, Nair Braga, Francisca Rodrigues * and M. Beatriz P. P. Oliveira

LAQV/REQUIMTE (Rede de Química e Tecnologia), Department of Chemical Sciences, Faculty of Pharmacy, University of Porto, Rua de Jorge Viterbo Ferreira, 228, 4050-313 Porto, Portugal; dianaandreiapinto@gmail.com (D.P.); nairbraga@gmail.com (N.B.); beatoliv@ff.up.pt (M.B.P.P.O.)
* Correspondence: franciscapintolisboa@gmail.com or fsarmento@ff.up.pt; Tel.: +351-220428500; Fax: +351-226093390

Received: 25 October 2017; Accepted: 20 November 2017; Published: 24 November 2017

Abstract: *Castanea sativa* fruit processing generates high amounts of by-products, mostly bur. Currently, the cosmetic industry has a great interest in natural extracts as antioxidant sources. In the present study, *C. sativa* bur extract was used as the active ingredient, in different amounts, in topical hydrogels. The formulations were characterized regarding total phenolic and flavonoid contents (TPC and TFC, respectively), antioxidant activity (DPPH radical scavenging capacity and ferric reducing antioxidant power (FRAP)) and technological and microbiological properties. The same parameters were evaluated after 30 days of storage at 4 °C ($T_{30/4}$ °C) and 20 °C ($T_{30/20}$ °C). At time 0 (T_0), the TPC ranged between 0.79 and 9.65 mg of gallic acid equivalents (GAE)/g gel, while TFC varied from 0.05 to 1.23 mg of catechin equivalents (CAE)/g gel. Antioxidant activity was high for both assays, with values at T_0 ranging between 98.41 and 1013.43 μmol of ferrous sulphate equivalents (FSE)/g gel and varying between 431.96 and 990.84 μg of Trolox equivalents (TE)/g gel for FRAP and DPPH assays, respectively. No formulation exceeded the defined criteria in microbiological counts. All formulations showed similar technological profiles but particular attention should be given to pH. The gel with 50% of extract (F3) was selected as the best one for potential cosmetic applications.

Keywords: *Castanea sativa* bur; chestnut by-product; antioxidants; skin-oxidative stress; hydrogels

1. Introduction

Castanea sativa Mill., also known as "sweet chestnut", is a species of the Fagaceae family common to south Europe and Asia (mainly in China). Chestnut is comprised of fruit, integument (inner shell), pericarp (outer shell) and bur. The high economic value of chestnut is due to their nutritional composition being largely appreciated in Mediterranean countries [1,2]. High amounts of by-products are generated by chestnut processing chains, mainly shell and bur. The waste residues obtained from food, forest or agricultural industries are nowadays considered attractive, since they are inexpensive and their re-use provides huge environmental benefits.

Recently, some studies have reported the potentialities of *C. sativa* by-products, such as leaf, shell and bur, disclosing its richness in antioxidant constituents. The content of phenolic compounds and the antioxidant activity are associated with beneficial health effects such as anticarcinogenic, anti-inflammatory, cardioprotective and neuroprotective activities [1,3–9]. Phenolic acids (ellagic and gallic acid), flavonoids (rutin, quercetin and apigenin) and tannins were the main phenolic compounds reported [5,10]. Nevertheless, *C. sativa* bur had no described application and usually remains in the woodland after fruit harvesting, promoting the insect larvae proliferation [5]. The valorisation of this waste represents a new challenge for the chestnut chain in order to encourage the research of novel applications, focusing on sustainable approaches and obtaining new valuable products [2,11].

Taking into consideration the potentialities shown by the bur, it is expected that this by-product, after specific pre-treatments, could be a source of added value natural compounds with high interest to pharmaceutical, food and cosmetic industries [3,5,12].

Regarding the studies about the applicability of these by-products, chestnut leaf has been used for the prevention of diabetes and DNA injuries [13], antibacterial activity [14] and protection from skin-oxidative stress [9,15]. The shell could be used as a heavy metal adsorbent [16] and phenol substitute in adhesive formulations [17]. However, to the best of our knowledge, no reports on the use of this by-product for cosmetic formulations are available.

One of the major targets for oxidative stress is skin, since this organ is permanently exposed to external aggressions such as ultraviolet radiation (UV) and atmospheric pollution. Indeed, skin diseases, such as psoriasis, atopic dermatitis and cancer have been related with cutaneous oxidative stress, photoaging and photocarcinogenesis [18–20]. Direct evidences that UV radiation contributes to the production of reactive species in skin have been reported [21], justifying the use of antioxidants in skin-care products. Besides that, in recent years, consumers have shown a clear preference for natural products, particularly natural antioxidants. Thus, agro-industrial by-products, described as good sources of phenolic compounds, have been exploited to find new natural antioxidants with the purpose to replace synthetics one, which have been associated with more toxic side effects [4,22]. The abundance of phenolic compounds is generally related to a high antioxidant activity, providing in vivo protection against free radical damage and decreasing the risk of oxidative stress-related diseases such as degenerative diseases [23]. Oxidative stress occurs when there is an overproduction of reactive oxygen species (ROS), reactive nitrogen species (RNS) and reactive sulphur species (RSS) that interact and injure the structure and function of proteins, DNA and RNA molecules, lipids and sugars [24,25]. In particular, phenolic acids are mostly responsible for the prevention of oxidative processes [26], while flavonoids are able to activate antioxidant enzymes [27]. Likewise, *C. sativa* kernels have high amounts of phenols in their composition, particularly gallic acid and ellagic acid [28].

The aim of this study was to formulate hydrogels with *C. sativa* hydroalcoholic extracts in different concentrations (F2, F3 and F4) and to compare their behaviour with a formulation without extract (F1). The bioactive compounds were quantified by the determination of total phenolic content (TPC) and total flavonoids content (TFC) while the antioxidant activity was evaluated by DPPH radical-scavenging ability and ferric reducing antioxidant power (FRAP) assays. Also, all formulations were characterized regarding microbiological quality and technological properties (namely, colour, pH, moisture, rheology and texture).

2. Materials and Methods

2.1. Chemicals and Reagents

1,1-diphenyl-2-picrylhydrazyl free radical, catechin, Folin–Ciocalteu's reagent, gallic acid, 6-hydroxy-2,5,7,8-tetramethylchroman-2-carboxylic acid (Trolox, a water-soluble derivative of vitamin E), 2,4,6-Tris(2-pyridyl)-s-triazine (TPTZ), ferrous sulphate heptahydrate and ferric chloride hexahydrate were all purchased from Sigma-Aldrich (Steinheim, Germany). Absolute ethanol, dimethylsulfoxide, sodium acetate, sodium carbonate decahydrate, sodium nitrite, aluminium chloride and sodium hydroxide were purchased from Merck (Darmstadt, Germany). All chemicals and solvents were of analytical grade. Aqueous solutions were prepared with deionized water from a Mili-Q water purification system (TGI Pure Water Systems, St Lincoln, NE, USA). Glycerine was obtained from Fluka (Steinheim, Germany). Carbopol® 940, polysorbate 80 and triethanolamine were purchased from Vaz Pereira (Lisbon, Portugal). Buffered peptone water, Tryptic Soy Agar medium, Nutrient Agar medium and Sabouraud Dextrose Agar medium were obtained from Oxoid Ltd. (Hampshire, UK).

2.2. Plant Materials

C. sativa burs were randomly collected in Trás-os-Montes, a north region of Portugal, after harvesting the chestnut. Bur was milled to a particle size of about 0.1 mm using an A11 basic analysis mill (IKAWearke, Staufen, Germany) and the fine powdered bur was stored in plastic tubes at 4 °C until the preparation of the extracts.

2.3. Preparation of C. sativa *Bur Hydro-Alcoholic Extracts*

One gram of powdered samples was submitted to solvent extraction by maceration at 40 °C for 30 min with 20 mL of ethanol: water (1:1 *v/v*). Hydro-alcoholic extracts were filtered through Whatman No. 1 filter paper and collected. Extracts were stored under refrigeration (4 °C) until use [3].

2.4. Preparation of Hydrogels Containing C. sativa *Bur Hydro-Alcoholic Extracts*

The first step was to dissolve glycerine and preservative in water, at room temperature. Glycerine provides emollient and moisturizing properties to the gel. Afterwards, *C. sativa* bur hydro-alcoholic extract was also dissolved in the previous mixture. Then, Carbopol® 940, acting as thickening and gelling agent, was slowly dispersed under magnetic stirring (750 rpm) at 40 °C during 1 h. After the formulation had cooled down, the mixture was homogenised and the triethanolamine added dropwise to complete the gelation process. The composition of hydrogel is detailed in Table 1.

Table 1. Composition of the hydrogels.

Compounds	Composition of Hydrogels			
	Total Weight (%)			
	F1	F2	F3	F4
Glycerine	10	10	10	10
Carbopol® 940	0.5	0.5	0.5	0.5
Preservative	0.5	0.5	0.5	0.5
C. sativa bur hydro-alcoholic extract	-	25	50	89
Triethanolamine	q.s.	q.s.	q.s	q.s.
Deionized water	89	64	39	-

q.s.—quantum satis.

The homogeneous appearance is conferred by slow shaking, avoiding air incorporation. The gels were packed in polyethylene jars and rested 24 h before any measurement. Three formulations with different concentrations of bur hydro-alcoholic extract (F2, F3 and F4) were prepared. Simultaneously, a hydrogel without extract (F1) was also prepared. This procedure was done twice to prepare two sets of formulations: one set was kept at 4 °C and the other at 20 °C. Each formulation was prepared in triplicate.

2.5. Technological Characterization of Hydrogels

2.5.1. Colour Evaluation

Colour was evaluated using a colorimeter (Chroma Meter CR-410, Konica Minolta, Tokyo, Japan) previously calibrated with a white reference background. An aliquot of each gel was placed on a Petri plate with 5.5 cm diameter and a sample thickness of 5 mm. The results were expressed according to the colour space CIE 1976 $L^* a^* b^*$, defined by the Comission Internationale de l'Éclairage (CIE). The three coordinates represent the lightness of the colour (L^*), its position between red and green (a^*) and its position between yellow and blue (b^*). For each sample, the results were expressed as the average of the three measurements of all parameters evaluated.

2.5.2. Determination of pH

pH was determined using a pHmeter (Basic 20+, Crison, Barcelona, Spain) equipped with a glass electrode previously calibrated. The measurements were performed in triplicate for each sample.

2.5.3. Determination of Moisture

The moisture content was determined using a moisture analyser (Scaltec model SMO01, Scaltec Instruments, Göttinge, Germany). About 1 g of gel was subjected to a drying process at 100 ± 2 °C for 5 min. This procedure was performed in triplicate and expressed as percentage.

2.5.4. Texture Analysis

Texture analysis was carried out under compression in a texturometer (Stable Micro Systems TA-XT2i, Godalming, UK) by performing a spreadability test at 20 °C using a penetration test probe of 23 mm. The probe applied a force of 0.049 N with a penetration depth of 3 mm and speed of 3 mm/s. The female cone loaded the gels samples, while the corresponding male cone came down towards the sample, forcing it upwards and outwards at 45°. Data were collected using the Exponent Stable Micro Systems software® (Stable Micro Systems Ltd, Godalming, UK).

2.5.5. Rheological Behaviour Analysis

Rheological analysis was carried out on a rotary viscometer (Brookfield Digital Model DV-E, Cologne, Germany). This procedure consists in the determination of the force exerted on a body fixed in rotation on a liquid at a constant speed of rotation. Flow measurements were performed at 20 ± 0.5 °C in the shear rate range from 0.1 to 500 s^{-1}, with 60 s delay period between measurements, and the speeds range evaluated was fixed between 5 and 50 rpm. Until performing the determinations, gels were kept resting for 30 min. The results were represented in a graph which relates the apparent viscosity, expressed as mPa·s, and the shear rate, expressed as rpm, for each gel analysed at different times and temperatures.

2.6. Determination of Hydrogels Bioactive Compounds

2.6.1. Total Phenolic Content

Total phenolic content (TPC) was spectrophotometrically determined according to the Folin-Ciocalteu procedure [29] with minor modifications. This methodology was directly performed in a 96 wells microplate based on a complex redox reaction between the phenolic compounds and the phosphotungstic and phosphomolybdic acids present in the Folin–Ciocalteu's reagent. All gel samples (100 mg) were previously diluted in 1 mL of dimethylsulfoxide (DMSO): ethanol (1:1 v/v). Briefly, the reaction mixture in each well consisted of 30 µL of diluted gel sample (100 mg/mL), 150 µL of Folin–Ciocalteu's reagent ($10 \times$ dilution) and 120 µL of Na_2CO_3 7.5% (w/v) solution. The microplate was incubated at 45 °C for 15 min in a Synergy HT Microplate Reader (BioTek Instruments, Inc., Winooski, VT, USA) and after left to stand in the dark for 30 min, at room temperature. The absorbance was read at 765 nm using the microplate reader described above. A calibration curve with gallic acid was used as reference (linearity range = 20–1000 µg/mL, $R^2 > 0.993$) to obtain a correlation between absorbance and standard concentration. The results were expressed as mg of gallic acid equivalents (GAE) per gram of gel (mg GAE/g gel). The assay was carried out in triplicate.

2.6.2. Total Flavonoid Content

Total flavonoid content (TFC) was performed by a colorimetric method based on the formation of flavonoid–aluminium compounds, according to the method described by Costa et al. [30], with minor modifications. This assay was directly carried out in a 96-well microplate. Gel samples (100 mg) were previously diluted in 1 mL of DMSO: ethanol (1:1 v/v). Briefly, in each well, 30 µL of diluted gel sample (100 mg/mL) was mixed with 75 µL of ultrapure water and 45 µL of $NaNO_2$ 1% (w/v) solution. After 5 min, 45 µL of $AlCl_3$ 5% (w/v) solution was added. Then, 1 min later, 60 µL of NaOH 1 mol/L and 45 µL of ultrapure water were also added. The absorbance was read at 510 nm using a Synergy HT Microplate Reader (BioTek Instruments, Inc., Winooski, VT, USA). In order to obtain a correlation between standard concentration and absorbance, a calibration curve was prepared with catechin (linearity range = 2.5–500 µg/mL, $R^2 > 0.996$). The results were expressed as mg of catechin equivalents (CAE) per gram of gel (mg CAE/g gel). This parameter was evaluated in triplicate.

2.7. In Vitro Antioxidant Activity of Hydrogels

Regarding the antioxidant activity, gels were evaluated by DPPH radical scavenging activity and ferric reducing antioxidant power (FRAP), as described in the following points.

2.7.1. DPPH Free Radical Scavenging Assay

The methodology principle is based on the evaluation of the antiradical activity of extracts by reduction of $DPPH^•$ to hydrazine. DPPH is a stable free radical with a maximum absorption between 515 and 517 nm. All gel samples (100 mg) were previously diluted in 1 mL of DMSO:ethanol (1:1 v/v). The reaction mixture was directly made in a 96-well microplate. 30 µL of diluted gel sample (100 mg/mL) and 270 µL of ethanol solution containing DPPH radicals in a final concentration of 6×10^5 M were put in each well. The microplate was left to stand at room temperature for 40 min, in the dark. Afterwards, the absorbance was measured at 517 nm using a Synergy HT Microplate Reader (BioTek Instruments, Inc., Winooski, VT, USA) to determine the DPPH radical reduction [31] in which the low absorbance values correspond to high free radical scavenging activity. A standard curve was plot with Trolox, used as reference (linearity range: 2.5–125 µg/mL, $R^2 > 0.970$). The antioxidant potential based on the DPPH free radical scavenging ability of gels samples was expressed as µg of Trolox equivalents (TE) per gram of gel (µg TE/g gel) of at least three determinations.

2.7.2. Ferric Reducing Antioxidant Power Assay

According to the procedure described by Benzie et al. [32] with minor modifications, the ferric reducing antioxidant power (FRAP) assay measures the reduction of a ferric 2,4,6-tripyridyl-s-triazine complex (Fe^{3+}-TPTZ) to the ferrous form (Fe^{2+}-TPTZ) in the presence of antioxidants. The chemical reaction involves a single electron reaction between $Fe(TPTZ)_2$ (III) and an electron donor ArOH (chemical structure presents in phenolic compounds). FRAP was directly performed in a 96-well microplate. About 100 mg of each gel sample was diluted in 1 mL of DMSO:ethanol (1:1 v/v). Briefly, the reaction mixture in each well consisted of 35 µL of diluted gel sample (100 mg/mL) and 265 µL of FRAP reagent (composed by 10 parts of 300 mM sodium acetate buffer at pH 3.6, 1 part of 10 mM TPTZ solution and 1 part of 20 mM $FeCl_3 \cdot 6H_2O$ solution). The microplate was incubated at 37 °C for 30 min. Then, the increase of absorbance was measured at 595 nm using a Synergy HT Microplate Reader (BioTek Instruments, Inc., Winooski, VT, USA). A calibration curve was prepared with ferrous sulphate ($FeSO_4 \cdot 7H_2O$) as standard (linearity range: 25–500 µM, $R^2 > 0.985$). The results were expressed as µmol of ferrous sulphate equivalents (FSE) per gram of gel (µmol FSE/g gel). This methodology was performed in triplicate.

2.8. Microbiological Properties

All formulations were evaluated regarding their total aerobic bacteria, mesophilic aerobic bacteria, and yeast and mould contents. Firstly, dilutions of each gel sample with buffered peptone sterile solution were prepared. 3% w/v of polysorbate 80 was added to neutralize the preservative function. After that, the mixture was homogenised and spread on the suitable medium. Total aerobic bacteria count was carried out after 5 days of incubation at 30–35 °C using tryptic soy agar medium, while mesophilic aerobic bacteria used nutrient agar medium, after incubation at 32.5 °C. For the yeast and molds count, Sabouraud dextrose agar medium was used and the plates incubated at 22.5 °C for 5 days. The results were expressed as number of colony-forming units per mL of gel dilution (CFU/mL). Each assay was performed in triplicate.

2.9. Stability Study

All formulations were evaluated at different storage times (T_0 and after 30 days (T_{30})) and different temperatures (4 and 20 °C), according to the protocols described previously.

2.10. Statistical Analysis

Data were reported as mean ± standard deviation of at least triplicate experiments. Statistical analysis of the results was performed with SPSS 22.0 (SPSS Inc., Chicago, IL, USA). One-way ANOVA was used to investigate the differences between samples for all assays. Post hoc comparisons of the means were performed according to Tukey's HSD test. In all cases, $p < 0.05$ was accepted as denoting significance.

3. Results and Discussion

3.1. Technological Characterization of Hydrogels

3.1.1. Colour Evaluation

Colour evaluation is a complementary methodology used to assess the acceptance of the product by consumers. The results of colour evaluation are summarized in Table 2.

Regarding L*, all gels had a considerable high lightness at T_0 (10.69–11.25). In what concerns a*, gels with 25% (F2) and 50% of extract (F3) exhibited positive values at T_0 (2.70 and 0.43, respectively) associated to the redness, while the gels without extract (F1) and with the highest extract amount (F4) showed a prevalence of greenness associated to the negative results (−0.64 and −0.28, respectively). Relatively to b*, at T_0, all formulations presented positive values associated to the prevalence of yellowness, being the highest results obtained in F2 and the lowest in F4 (6.39 and 1.69, respectively). F4 showed the highest lightness and the lowest prevalence of yellowness, being in agreement with the dark colour detected by the eye.

3.1.2. Determination of pH

Skin-care formulations must present a pH compatible with skin (between 5 and 6) [33]. The results at T_0 are present in Table 2. The highest pH values were obtained for gels with the highest extract percentage. Therefore, the addition of bur hydro-alcoholic extract gives a basic pH to the formulations. F4 revealed the highest pH, with values greater than 8, which is incompatible with skin application. On the contrary, F1 had the lowest pH with a value minor than 5, which is also unsuitable to scattering on the skin. Due to the skin compatibility issue, F2 and F3 were selected as the best ones, with pH values of 5.27 and 5.67, respectively, which are more suitable for cutaneous application.

Table 2. Colour variation (L^*, a^* and b^*), pH, moisture and texture (adhesiveness and firmness) of hydrogels containing chestnut bur extracts in different percentages, evaluated at different times and temperatures (mean ± standard deviation, $n = 3$). T_0, time 0. $T_{30/4}$ °C, after 30 days at 4 °C. $T_{30/20}$ °C, after 30 days at 20 °C.

Formulations		L* (Lightness)	a* (Redness)	b* (Yellowness)	pH	Moisture (%)	Adhesiveness (N.mm)	Firmness (N)
					Technological Properties			
F1	T_0	10.69 ± 0.01 [a]	−0.64 ± 0.04 [a]	3.37 ± 0.02 [c]	4.46 ± 0.04 [a]	30.69 ± 1.64 [a]	−1.480 ± 0.237 [a]	0.554 ± 0.013 [b]
	$T_{30/4}$ °C	10.01 ± 0.05 [c]	−0.72 ± 0.06 [a]	4.01 ± 0.04 [a]	4.48 ± 0.01 [a]	30.28 ± 0.70 [a]	−1.943 ± 0.232 [a]	0.749 ± 0.023 [a]
	$T_{30/20}$ °C	10.11 ± 0.01 [b]	−0.73 ± 0.04 [a]	3.57 ± 0.02 [b]	4.52 ± 0.01 [a]	29.35 ± 0.55 [a]	−1.413 ± 0.220 [a]	0.531 ± 0.031 [b]
F2	T_0	10.74 ± 0.02 [c]	2.70 ± 0.08 [a]	6.39 ± 0.06 [b]	5.27 ± 0.03 [a]	20.27 ± 0.39 [a]	−1.158 ± 0.148 [a]	0.461 ± 0.020 [a]
	$T_{30/4}$ °C	10.93 ± 0.01 [a]	2.65 ± 0.04 [a]	6.71 ± 0.06 [a]	5.00 ± 0.02 [b]	20.01 ± 0.66 [a]	−1.193 ± 0.013 [a]	0.471 ± 0.019 [a]
	$T_{30/20}$ °C	10.83 ± 0.02 [b]	2.74 ± 0.08 [a]	6.51 ± 0.06 [b]	4.99 ± 0.01 [b]	20.12 ± 0.49 [a]	−1.029 ± 0.077 [a]	0.429 ± 0.004 [b]
F3	T_0	10.31 ± 0.02 [b]	0.43 ± 0.03 [a]	3.03 ± 0.01 [a]	5.67 ± 0.01 [b]	15.43 ± 3.13 [a]	−0.932 ± 0.122 [a]	0.367 ± 0.012 [b]
	$T_{30/4}$ °C	10.96 ± 0.02 [a]	0.42 ± 0.02 [a]	2.89 ± 0.02 [c]	5.65 ± 0.01 [b]	14.15 ± 0.53 [a]	−1.125 ± 0.033 [b]	0.439 ± 0.011 [a]
	$T_{30/20}$ °C	10.27 ± 0.02 [b]	0.49 ± 0.03 [a]	2.98 ± 0.01 [b]	5.72 ± 0.01 [a]	13.89 ± 0.73 [a]	−0.857 ± 0.035 [b]	0.381 ± 0.005 [b]
F4	T_0	11.25 ± 0.02 [c]	−0.28 ± 0.09 [a]	1.69 ± 0.01 [c]	8.28 ± 0.01 [b]	14.04 ± 0.63 [a]	−0.707 ± 0.082 [a]	0.288 ± 0.008 [a]
	$T_{30/4}$ °C	12.44 ± 0.03 [a]	−0.15 ± 0.09 [a]	1.96 ± 0.05 [a]	8.69 ± 0.01 [a]	14.03 ± 0.13 [a]	−0.659 ± 0.029 [a]	0.259 ± 0.007 [b]
	$T_{30/20}$ °C	11.32 ± 0.02 [b]	−0.35 ± 0.09 [c]	1.78 ± 0.01 [b]	8.21 ± 0.02 [c]	13.99 ± 0.23 [a]	−0.679 ± 0.060 [a]	0.261 ± 0.007 [b]

Different letters (a, b, c) in the same line of the same column indicate significant differences ($p < 0.05$) between the results obtained during gels storage.

3.1.3. Determination of Moisture

Some constituents of cosmetic preparations, such as water, usually provide an appropriate medium for microbial growth and could stimulate physical and chemical changes in skin-care products. The moisture results (Table 2) show that at T_0, F1 has the highest moisture content (30.69%) while F3 and F4 displayed the lowest values (15.43% and 14.04%, respectively). The increase of the concentration of *C. sativa* bur extract could decrease the possibility of microbial growth and physical and chemical modifications due to the reduction of moisture content.

3.1.4. Texture Analysis

An appropriate texture is one of the principal properties that cosmetics should present to improve consumer acceptance. This methodology evaluates firmness and adhesiveness (Table 2). According to the results, it is possible to conclude that adhesiveness increases with the presence of bur extract. At T_0, F1 showed the lowest adhesiveness with −1.480 N·mm, while the highest result was obtained for F4 (−0.707 N·mm). In opposite to adhesiveness, the increase of extract concentration in gels affords the decrease of firmness. At T_0, F1 presented the highest firmness (0.554 N), while F4 had the lowest (0.288 N).

3.1.5. Rheological Behaviour Analysis

Figure 1 represents the apparent viscosity with the shear rate in formulations at described conditions.

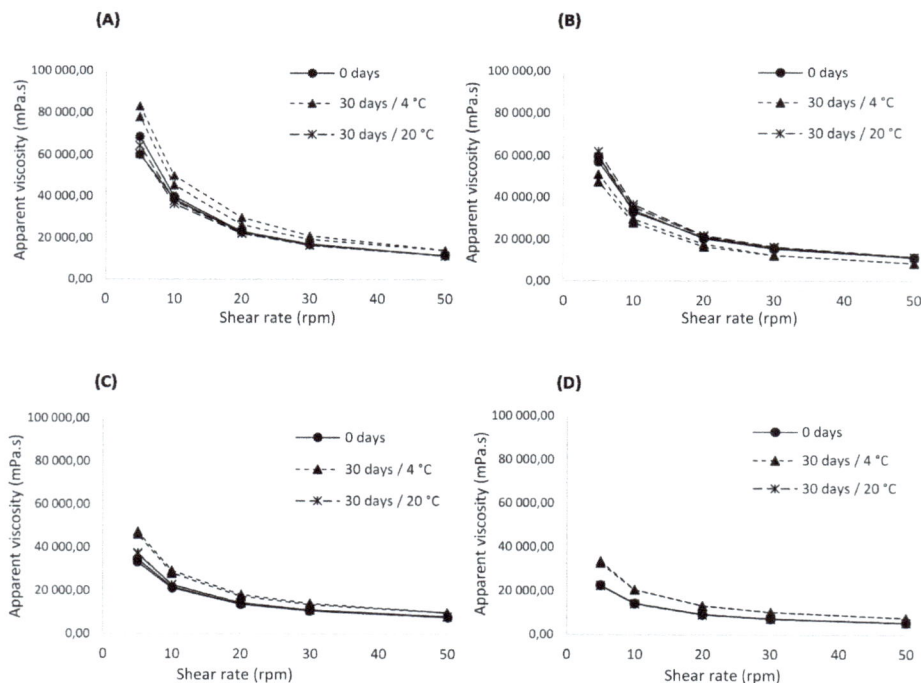

Figure 1. Comparison of the rheograms of hydrogels containing chestnut bur extracts in different percentages, evaluated at T_0, $T_{30/4 \,°C}$ and $T_{30/20 \,°C}$. Values are an average of three individual experiments ($n = 3$) expressed as mean ± standard deviation. (**A**) Formulation without extract (F1); (**B**) Formulation with 25% of extract (F2); (**C**) Formulation with 50% of extract (F3); (**D**) Formulation with the highest extract amount (F4).

In what concerns the apparent viscosity, F1 showed the highest viscosity values while F4 showed the lowest ones. Therefore, apparent viscosity decreases with increasing bur extract percentages in hydrogels. All formulations presented a rheofluidificant behaviour since apparent viscosity decreased with the increasing of shear rate (Figure 1). This rheological behaviour is characteristic of water in oil emulsions, and is similar to gels prepared with acrylic polymers [34]. A high viscosity of the semisolid formulation could have a positive influence on the stability of extract once it decreases the diffusion rate of oxygen [35].

3.2. Bioactive Compounds in Hydrogels

3.2.1. Total Phenolic Content

The Folin–Ciocalteu's phenol reagent is frequently used to estimate the amount of phenolic compounds in herbal extracts. This is a non-specific assay once the phenol reagent could react not only with phenolic compounds but also with other substances (interfering substances) such as ascorbic acid, aromatic amines, sugars and sulphur dioxide [4]. Results of this parameter are displayed in Table 3.

Table 3. Total phenolic content (TPC), total flavonoid content (TFC), ferric reducing antioxidant power (FRAP) and 2,2-diphenyl-1-picrylhydrazyl (DPPH•) scavenging assay of hydrogels containing chestnut bur extracts in different percentages, evaluated at different times and temperatures (mean ± standard deviation, n = 3). T_0, time 0. $T_{30/4\ °C}$, after 30 days at 4 °C. $T_{30/20\ °C}$, after 30 days at 20 °C. GAE, gallic acid equivalents. CAE, catechin equivalents. FSE, ferrous sulphate equivalents. TE, Trolox equivalents.

Formulations		Bioactive Compounds			
		TPC (mg GAE/g gel)	TFC (mg CAE/g gel)	FRAP (μmol FSE/g gel)	DPPH (μg TE/g gel)
F1	T_0	0.79 ± 0.01 [a]	0.05 ± 0.01 [a]	98.41 ± 7.65 [a]	431.96 ± 7.71 [a]
	$T_{30/4\ °C}$	0.82 ± 0.01 [a]	0.06 ± 0.01 [a]	85.05 ± 5.01 [a]	395.01 ± 3.86 [b]
	$T_{30/20\ °C}$	0.69 ± 0.01 [b]	0.06 ± 0.01 [a]	76.71 ± 5.78 [b]	280.46 ± 8.12 [c]
F2	T_0	2.48 ± 0.02 [b]	0.31 ± 0.01 [a]	509.17 ± 16.10 [a]	658.37 ± 5.26 [a]
	$T_{30/4\ °C}$	2.81 ± 0.01 [a]	0.27 ± 0.01 [b]	510.84 ± 13.25 [a]	656.07 ± 10.11 [a]
	$T_{30/20\ °C}$	1.96 ± 0.01 [c]	0.23 ± 0.01 [c]	412.32 ± 16.10 [b]	611.24 ± 10.21 [b]
F3	T_0	5.30 ± 0.01 [a]	0.73 ± 0.01 [a]	774.66 ± 10.43 [a]	803.15 ± 12.71 [a]
	$T_{30/4\ °C}$	4.41 ± 0.02 [b]	0.69 ± 0.01 [b]	727.91 ± 15.30 [b]	795.03 ± 7.58 [a]
	$T_{30/20\ °C}$	3.48 ± 0.01 [c]	0.63 ± 0.01 [c]	714.55 ± 10.43 [b]	745.91 ± 13.36 [b]
F4	T_0	9.65 ± 0.04 [a]	1.23 ± 0.02 [a]	1013.43 ± 12.61 [a]	990.84 ± 14.06 [a]
	$T_{30/4\ °C}$	8.00 ± 0.01 [b]	1.16 ± 0.01 [b]	1030.13 ± 18.97 [a]	933.98 ± 11.58 [b]
	$T_{30/20\ °C}$	7.55 ± 0.04 [c]	1.08 ± 0.02 [c]	888.20 ± 12.61 [b]	867.11 ± 11.57 [c]

Different letters (a, b, c) in the same line of the same column indicate significant differences ($p < 0.05$) between the results obtained during gels storage.

The gels TPC at T_0 ranged from 0.79 (F1) to 9.65 mg (F4) of GAE/g of gel. As expected, the amount of phenolic compounds strongly increases with the concentration of chestnut bur extract. For this reason, gels with higher percentages of bur extract that showed greater TPC values should also present higher antioxidant activity. The incorporation of chestnut bur in a dermatological basis still presents high TPC, in the same range of some plant by-products, highlighting its potential source of polyphenols [7,8,23].

3.2.2. Total Flavonoid Content

Flavonoids represent a group from phenolic compounds broadly spread in a wide variety of plants. As a group belonging to phenolic compounds, flavonoids are also described as contributors to the antioxidant activity (Table 3). Gels at T_0 showed a TFC ranging from 0.05 to 1.23 mg CAE/g of gel. Similar to TPC, F4 presented the highest TFC while F1 has the lowest value.

3.3. In Vitro Antioxidant Activity of Hydrogels

3.3.1. DPPH Free Radical Scavenging Assay

DPPH scavenging capacity is a method widely used to determine antioxidant activity in plant and fruit extracts [36]. The methodology principle is based on the reduction of DPPH radicals in ethanol providing a decrease in absorbance at 517 nm. It was decided to express the results as μg of TE per g of gel as it is more meaningful and provides a better interpretation than the results expressed as inhibition percentages, since the decrease in absorbance is not enough to achieve 50% of inhibition. The results (Table 3) show that, at T_0, the antioxidant capacity values range from 431.96 to 990.84 μg of TE/g of gel. The increasing order of antioxidant power for gels was: F1 < F2 < F3 < F4. As expected, the antioxidant activity based on DPPH radical scavenging ability increases with the extract concentration added to the formulation. According to Pinto et al., a concentration range between 38.67 and 76.86 μg/mL of *C. sativa* bur extract is responsible for at least 50% of inhibition, since the results were expressed as EC_{50}. Besides the bur extract, the formulation contains other compounds that could explain the lower antioxidant activity when compared with *C. sativa* bur extract [3].

3.3.2. Ferric Reducing Antioxidant Power Assay

Antioxidant activity evaluated through FRAP assays determines the ability of plant extracts to reduce ferric ions. Table 3 shows the antioxidant capacity averages obtained for each formulation. Similar to DPPH assays, the increasing order of antioxidant activity was: F1 < F2 < F3 < F4. As expected, at T_0, F4 showed the highest antioxidant activity by ferric reduction (1013.43 μmol of FSE/g gel), while F1 had the lowest reduction power (98.41 μmol of FSE/g gel). Compared to the *C. sativa* bur, the antioxidant power of hydrogels was slightly lower [3].

3.4. Correlation between Total Phenolic Content and Antioxidant Activity

As referred antioxidant activity has been evaluated based on DPPH and FRAP assays. The possible correlation between these results and the TPC was evaluated through Pearson correlation. This correlation coefficient measures the degree of linear correlation between two quantitative variables. When a correlation analysis was performed between TPC and FRAP or DPPH assays, a strong positive correlation was observed for FRAP ($R^2 = 0.9360$) and DPPH ($R^2 = 0.9233$), inferring a possible responsibility of TPC for the antioxidant activity observed. Nevertheless, these results could be influenced by synergic effects of the gels constituents or other chemical classes.

3.5. Correlation between Total Flavonoid Content and Antioxidant Activity

When a correlation analysis by Pearson coefficient was performed between TFC and FRAP or DPPH assays, an extremely positive correlation was also observed for both methods with values of $R^2 = 0.9572$ and $R^2 = 0.9320$, respectively. Further studies, such as HPLC-MS, would be helpful to identify the flavonoid compounds that contribute to this correlation.

3.6. Correlation between Total Phenolic and Total Flavonoid Contents

Analysing the correlation through the Pearson coefficient, an extremely positive correlation ($R^2 = 0.9848$) was established between TPC and TFC.

3.7. Microbiological Properties

The microbiological results are summarized in Table 4, being expressed as colony forming units per mL of gel (CFU/mL). Concerning the three groups of microorganisms analysed, all formulations presented values lower than the limits defined by ISO standard 11930:2012 (the number of colony building units (CFU) on solid culture media should be inferior to 1000 CFU/mL), demonstrating the preservative efficacy at the storage conditions used [37].

Table 4. Total aerobic, total mesophilic aerobic bacteria and total yeast and mould counts of hydrogels containing chestnut bur extracts in different percentages, evaluated at T_0, $T_{30/4}$ °C and $T_{30/20}$ °C. T_0, time 0. $T_{30/4}$ °C, after 30 days at 4 °C. $T_{30/20}$ °C, after 30 days at 20 °C. Values are expressed as mean of the replicates ($n = 3$) in colony forming units per mL of gel (CFU/mL gel).

Formulations		Microbiological Properties		
		Total Aerobic Count (UFC/mL gel)	Total Mesophilic Aerobic Bacteria Count (UFC/mL gel)	Total Yeast and Mold Count (UFC/mL gel)
F1	T_0	≤ 10	≤ 10	≤ 10
	$T_{30/4}$ °C	≤ 10	≤ 10	≤ 10
	$T_{30/20}$ °C	≤ 10	≤ 10	≤ 10
F2	T_0	≤ 10	≤ 10	≤ 10
	$T_{30/4}$ °C	≤ 10	≤ 10	≤ 10
	$T_{30/20}$ °C	≤ 10	≤ 10	≤ 10
F3	T_0	≤ 10	≤ 10	195
	$T_{30/4}$ °C	≤ 10	≤ 10	≤ 10
	$T_{30/20}$ °C	≤ 10	≤ 10	≤ 10
F4	T_0	≤ 10	15	15
	$T_{30/4}$ °C	≤ 10	25	30
	$T_{30/20}$ °C	≤ 10	≤ 10	≤ 10

3.8. Stability Study

As previously mentioned, all technological parameters were evaluated after 30 days of storage at 4 °C and 20 °C. In what concerns to L* (lightness), the results for each formulation were statistically different ($p < 0.05$) at T_0, $T_{30/4}$ °C and $T_{30/20}$ °C, except in F3. All gels presented high lightness values. Regarding a* coordinates, F2 and F3 at $T_{30/4}$ °C and $T_{30/20}$ °C maintained the prevalence of redness, while F1 and F4 still presented greenness as prevalent. This parameter did not seem to be related to the storage time and temperatures since there are no significant differences ($p > 0.05$), except for F4 that displayed a significant decrease in gels at $T_{30/20}$ °C ($p < 0.05$). Relative to b* parameter, similar to T_0, all gels at $T_{30/4}$ °C and $T_{30/20}$ °C also presented a prevalence of yellowness. The b* results for each formulation at different storage conditions were significantly different ($p < 0.05$).

The pH analysis after 30 days of storage at both conditions revealed similar results to the gels at T_0. The pH of F4 and F1 were also not compatible with skin application. The pH values of F1 were not statistically different ($p > 0.05$) at T_0 and after 30 days.

Considering the moisture content, gels analysed at different times and temperatures showed a decrease of moisture with the increase of bur extract concentration in hydrogels. All formulations at T_0 and after 30 days did not show significant differences ($p > 0.05$) in moisture percentages. For this reason, storage time and temperatures did not affect the moisture contents, emphasizing the gel stability.

Similar to T_0, at $T_{30/4}$ °C and $T_{30/20}$ °C, adhesiveness increases and firmness decreases with the extract concentration in hydrogels. Until F3, the adhesiveness was lower at $T_{30/4}$ °C and higher at $T_{30/20}$ °C, except for F4 which had a higher adhesiveness at $T_{30/4}$ °C and lower at T_0. In what concerns adhesiveness, the results for each formulation were not statistically different ($p > 0.05$) at T_0, $T_{30/4}$ °C and $T_{30/20}$ °C, except for F3 at $T_{30/4}$ °C ($p < 0.05$). In general, the adhesiveness of *C. sativa* bur hydrogels is not dependent on temperature and time. On the other hand, gel firmness also decreases with the extract after 30 days of storage at 4 °C and 20 °C. Nevertheless, gels up to 50% of extract at $T_{30/4}$ °C had the highest firmness, while the results at $T_{30/20}$ °C were the lowest ones.

A decrease in apparent viscosity in gels with higher bur extract concentrations was also found after 30 days of storage at different temperatures. At $T_{30/4}$ °C, F2 presented a viscosity lower than at T_0 and $T_{30/20}$ °C. However, for the other formulations, a higher viscosity was found at $T_{30/4}$ °C while the results at T_0 and $T_{30/20}$ °C were similar.

In what concerns TPC of gels, at $T_{30/4}$ °C and $T_{30/20}$ °C the highest contents were also displayed for F4. These formulations experienced significant changes in TPC with the highest contents obtained for gels at T_0. Indeed, the TPC of gels with extract declined after 30 days of storage, with the highest decrease at 20 °C. The results for F1 at T_0 and $T_{30/4}$ °C were not significantly different ($p > 0.05$), while all other formulations containing bur extract presented statistically different results ($p < 0.05$). Therefore, gel TPC was affected by storage time and temperature.

Relative to TFC, after storage time at different temperatures, F4 revealed the highest contents. Particularly, all gels with extract analysed at T_0 contained the highest amounts of flavonoids. After 30 days of storage, the TFC decreased in all formulations, with the highest decline determined in gels at $T_{30/20}$ °C. Results relating to F1 were not statistically different ($p > 0.05$). Besides the formulations with bur extract added had results significantly different ($p < 0.05$) at distinctive storage conditions. Thus, the time and temperature contributed to the TFC reduction on chestnut bur hydrogels.

Regarding the DPPH assay, all gels at T_0 showed the highest antioxidant activity, while the lowest was displayed at $T_{30/20}$ °C. Similar to the previous assays, gels with higher extract concentrations also presented the highest antioxidant activity based on DPPH radical scavenging capacity. The highest DPPH scavenging ability obtained at T_0 displayed significant differences ($p < 0.05$) for F1 and F4, whose antioxidant activity also presented a significant decrease ($p < 0.05$) at $T_{30/4}$ °C and $T_{30/20}$ °C. F2 and F3 showed a statistically significant decrease ($p < 0.05$) at $T_{30/20}$ °C.

Antioxidant activity evaluated through FRAP assays in gels analysed at $T_{30/4}$ °C and $T_{30/20}$ °C showed the highest ferric reduction power for F4, while F1 had the lowest results. At $T_{30/20}$ °C, the ferric reducing ability underwent the highest reduction in all formulations, presenting significant differences ($p < 0.05$) for F1, F2 and F4. In this case, time and temperature provided a decrease of antioxidant capacity. Significant differences ($p < 0.05$) for F3 were observed between T_0 and $T_{30/4}$ °C.

Microbiological analysis revealed similar results in all formulations along the storage time, evidencing microbiological stability. All gels after 30 days of storage presented results within the permissible limits and similar to gels at T_0. Furthermore, the storage temperatures did not seem to influence microbial growth since the counts at 4 and 20 °C are similar.

The incorporation of antioxidants extracted from plants in hydrogels represents a huge challenge since these compounds have been described as unstable in aqueous phases. There are many studies in this field reporting the physical instability of formulations with antioxidants added or the loss of extracts activity after incorporation in a cosmetic base [38,39]. In fact, the physical stability of semisolid formulations could be influenced by the chemical instability of some added constituents [40]. The high TPC, TFC and antioxidant activity of gels at T_0 means that *C. sativa* bur extract retained its functionality after being incorporated in the hydrogels. After 30 days, in formulations stored at 20 °C, there was a higher decrease in antioxidant activity compared with gels stored at 4 °C. The same is observed in TPC and TFC which could be related to the degradation of these compounds mediated by various factors such as storage aging, oxygen, aqueous phase of gel and temperature. Indeed, gels stored at 4 °C showed results similar or slightly lower than gels at T_0. Thus, it is possible to conclude that the lower temperature is better to prevent antioxidant degradation.

Besides the minor changes, taken together all the results it is possible to confirm a good physical, technological and microbiological stability of the formulations when stored during 30 days at 4 °C. However, the antioxidant properties showed changes more extensive after 30 days of storage. To prevent the decrease of antioxidant potential on gels provided by *C. sativa* bur extracts, the possible interferences with other constituents of gels should be studied and its composition should be optimized.

4. Conclusions and Future Perspectives

Nowadays, the cosmetic industry is a field on the rise, increasing the demand for novel ingredients, preferably from natural sources. For this reason, the use of *C. sativa* bur as a source of phytochemicals with beneficial health effects in skin-care formulations contributes to the advance of this industry and consequently to the environmental and economic sustainability. The valorisation impact of this by-product could be huge, taking into account the broad geographic distribution of this crop. Also, this study gives, for the first time, a comprehensive evaluation of the applicability of chestnut bur extract in skin-care formulations.

In this work, different percentages of bur hydro-alcoholic extracts were incorporated in a hydrogel base. There is a proportional ratio between the concentration of extract in the hydrogel and its TPC, TFC and antioxidant activity; thereafter F4 presented the highest values. However, its pH is basic and not compatible with skin pH. Considering all the results, F3 was selected as the best one for cosmetic purposes based on the high values of TPC, TFC and antioxidant activity, and the technological properties which are suitable for skin application and, consequently, enhance the consumer acceptance. Regarding gel composition, the preservative proved to be efficient in preventing the growth of microorganisms. However, the composition of hydrogels could be improved to increase consumer compliance. For example, the appearance might be upgraded by addition of an opacifier such as titanium dioxide. Also, triethanolamine could be replaced by citric acid which is a natural ingredient used in the adjustment of acid/base balance.

Further investigations are needed to assess the safety of gels including in vitro studies, such as cell viability and cytotoxicity in different skin cell lines, as well as skin and ocular irritability tests. Likewise, in vivo studies, such as patch tests and sensitization studies, must be carried out.

Acknowledgments: Francisca Rodrigues is thankful for her post-doc research grant from the project Operação NORTE-01-0145-FEDER-000011. Diana Pinto is thankful for the research grant from project UID/QUI/50006. This work received financial support from the European Union (FEDER funds through COMPETE), under the Partnership Agreement PT2020, and National Funds (FCT, Foundation for Science and Technology) through project LAQV/UID/QUI/50006/2013.

Author Contributions: Francisca Rodrigues conceived and designed the experiments; Diana Pinto performed the experiments; Francisca Rodrigues and Diana Pinto analysed the data; Beatriz Oliveira contributed reagents, materials and analysis tools; Diana Pinto and Beatriz Oliveira wrote the paper.

Conflicts of Interest: The authors declare no conflict of interest.

References

1. Rodrigues, F.; Santos, J.; Pimentel, F.B.; Braga, N.; Palmeira-de-Oliveira, A.; Oliveira, M.B.P.P. Promising new applications of *Castanea sativa* shell: Nutritional composition, antioxidant activity, amino acids and vitamin E profile. *Food Funct.* **2015**, *6*, 2854–2860. [CrossRef] [PubMed]
2. Braga, N.; Rodrigues, F.; Oliveira, M.B.P.P. *Castanea sativa* by-products: A review on added value and sustainable application. *Nat. Prod. Res.* **2015**, *29*, 1–18. [CrossRef] [PubMed]
3. Pinto, D.; Rodrigues, F.; Braga, N.; Santos, J.; Pimentel, F.B.; Palmeira-de-Oliveira, A.; Oliveira, M.B.P.P. The *Castanea sativa* bur as a new potential ingredient for nutraceutical and cosmetic outcomes: Preliminary studies. *Food Funct.* **2017**, *8*, 201–208. [CrossRef] [PubMed]
4. Balasundram, N.; Sundram, K.; Samman, S. Phenolic compounds in plants and agri-industrial by-products: Antioxidant activity, occurrence, and potential uses. *Food Chem.* **2006**, *99*, 191–203. [CrossRef]
5. Vázquez, G.; Fernández-Agulló, A.; Gómez-Castro, C.; Freire, M.S.; Antorrena, G.; González-Álvarez, J. Response surface optimization of antioxidants extraction from chestnut (*Castanea sativa*) bur. *Ind. Crops Prod.* **2012**, *35*, 126–134. [CrossRef]
6. Almeida, I.F.; Valentão, P.; Andrade, P.B.; Seabra, R.M.; Pereira, T.M.; Amaral, M.H.; Costa, P.C.; Bahia, M.F. In vivo skin irritation potential of a *Castanea sativa* (Chestnut) leaf extract, a putative natural antioxidant for topical application. *Basic Clin. Pharmacol. Toxicol.* **2008**, *103*, 461–467. [CrossRef] [PubMed]
7. Barreira, J.C.M.; Ferreira, I.C.F.R.; Oliveira, M.B.P.P.; Pereira, J.A. Antioxidant activities of the extracts from chestnut flower, leaf, skins and fruit. *Food Chem.* **2008**, *107*, 1106–1113. [CrossRef]
8. Barreira, J.C.M.; Ferreira, I.C.F.R.; Oliveira, M.B.P.P.; Pereira, J.A. Antioxidant Potential of Chestnut (*Castanea sativa* L.) and Almond (*Prunus dulcis* L.) By-products. *Food Sci. Technol. Int.* **2010**, *16*, 209–216. [CrossRef] [PubMed]

9. Almeida, I.F.; Maleckova, J.; Saffi, R.; Monteiro, H.; Góios, F.; Amaral, M.H.; Costa, P.C.; Garrido, J.; Silva, P.; Pestana, N.; et al. Characterization of an antioxidant surfactant-free topical formulation containing *Castanea sativa* leaf extract. *Drug Dev. Ind. Pharm.* **2015**, *41*, 148–155. [CrossRef] [PubMed]

10. Vasconcelos, M.C.B.M.; Bennett, R.N.; Quideau, S.; Jacquet, R.; Rosa, E.A.S.; Ferreira-Cardoso, J.V. Evaluating the potential of chestnut (*Castanea sativa Mill.*) fruit pericarp and integument as a source of tocopherols, pigments and polyphenols. *Ind. Crops Prod.* **2010**, *31*, 301–311. [CrossRef]

11. Federici, F.; Fava, F.; Kalogerakis, N.; Mantzavinos, D. Valorisation of agro-industrial by-products, effluents and waste: Concept, opportunities and the case of olive mill wastewaters. *J. Chem. Technol. Biotechnol.* **2009**, *84*, 895–900. [CrossRef]

12. Vázquez, G.; González-Alvarez, J.; Freire, M.S.; Fernández-Agulló, A.; Santos, J.; Antorrena, G. Chestnut Burs as a Source of Natural Antioxidants. In *Chemical Engineering Transactions*; United States Environmental Protection Agency (EPA): Washington, DC, USA, 2009; pp. 855–860.

13. Mujić, A.; Grdović, N.; Mujić, I.; Mihailović, M.; Živković, J.; Poznanović, G.; Vidaković, M. Antioxidative effects of phenolic extracts from chestnut leaves, catkins and spiny burs in streptozotocin-treated rat pancreatic β-cells. *Food Chem.* **2011**, *125*, 841–849. [CrossRef]

14. Basile, A.; Sorbo, S.; Giordano, S.; Ricciardi, L.; Ferrara, S.; Montesano, D.; Castaldo, R.C.; Vuotto, M.L.; Ferrara, L. Antibacterial and allelopathic activity of extract from *Castanea sativa* leaves. *Fitoterapia* **2000**, *71* (Suppl. 1), S110–S116. [CrossRef]

15. Almeida, I.F.; Costa, P.C.; Bahia, M.F. Evaluation of Functional Stability and Batch-to-Batch Reproducibility of a *Castanea sativa* Leaf Extract with Antioxidant Activity. *AAPS PharmSciTech* **2010**, *11*, 120–125. [CrossRef] [PubMed]

16. Vázquez, G.; Calvo, M.; Freire, M.S.; González-Álvarez, J.; Antorrena, G. Chestnut shell as heavy metal adsorbent: Optimization study of lead, copper and zinc cations removal. *J. Hazard. Mater.* **2009**, *172*, 1402–1414. [CrossRef] [PubMed]

17. Vázquez, G.; González-Alvarez, J.; Santos, J.; Freire, M.S.; Antorrena, G. Evaluation of potential applications for chestnut (*Castanea sativa*) shell and eucalyptus (*Eucalyptus globulus*) bark extracts. *Ind. Crops Prod.* **2009**, *29*, 364–370. [CrossRef]

18. Okayama, Y. Oxidative stress in allergic and inflammatory skin diseases. *Curr. Drug Targets Inflamm. Allergy* **2005**, *4*, 517–519. [CrossRef] [PubMed]

19. Sander, C.S.; Chang, H.; Hamm, F.; Elsner, P.; Thiele, J.J. Role of oxidative stress and the antioxidant network in cutaneous carcinogenesis. *Int. J. Dermatol.* **2004**, *43*, 326–335. [CrossRef] [PubMed]

20. Nishigori, C.; Hattori, Y.; Toyokuni, S. Role of reactive oxygen species in skin carcinogenesis. *Antioxid. Redox Signal.* **2004**, *6*, 561–570. [CrossRef] [PubMed]

21. Ou-Yang, H.; Stamatas, G.; Saliou, C.; Kollias, N. A chemiluminescence study of UVA-induced oxidative stress in human skin in vivo. *J. Investig. Dermatol.* **2004**, *122*, 1020–1029. [CrossRef] [PubMed]

22. Rechner, A.R.; Kuhnle, G.; Bremner, P.; Hubbard, G.P.; Moore, K.P.; Rice-Evans, C.A. The metabolic fate of dietary polyphenols in humans. *Free Radic. Biol. Med.* **2002**, *33*, 220–235. [CrossRef]

23. Rodrigues, F.; Palmeira-de-Oliveira, A.; Neves, J.; Sarmento, B.; Amaral, M.H.; Oliveira, M.B.P.P. Coffee silverskin: A possible valuable cosmetic ingredient. *Pharm. Biol.* **2015**, *53*, 386–394. [CrossRef] [PubMed]

24. Lu, J.M.; Lin, P.H.; Yao, Q.; Chen, C. Chemical and molecular mechanisms of antioxidants: Experimental approaches and model systems. *J. Cell Mol. Med.* **2010**, *14*, 840–860. [CrossRef] [PubMed]

25. Craft, B.D.; Kerrihard, A.L.; Amarowicz, R.; Pegg, R.B. Phenol-Based Antioxidants and the in vitro Methods Used for Their Assessment. *Compr. Rev. Food Sci. Food Saf.* **2012**, *11*, 148–173. [CrossRef]

26. Roleira, F.M.; Siquet, C.; Orru, E.; Garrido, E.M.; Garrido, J.; Milhazes, N.; Podda, G.; Paiva-Martins, F.; Reis, S.; Carvalho, R.A.; et al. Lipophilic phenolic antioxidants: Correlation between antioxidant profile, partition coefficients and redox properties. *Biorgan. Med. Chem.* **2010**, *18*, 5816–5825. [CrossRef] [PubMed]

27. Prochazkova, D.; Bousova, I.; Wilhelmova, N. Antioxidant and prooxidant properties of flavonoids. *Fitoterapia* **2011**, *82*, 513–523. [CrossRef] [PubMed]

28. Vasconcelos, M.C.B.M.; Bennett, R.N.; Rosa, E.A.S.; Ferreira-Cardoso, J.V. Primary and Secondary Metabolite Composition of Kernels from Three Cultivars of Portuguese Chestnut (*Castanea sativa Mill.*) at Different Stages of Industrial Transformation. *J. Agric. Food Chem.* **2007**, *55*, 3508–3516. [CrossRef] [PubMed]

29. Alves, R.C.; Costa, A.S.G.; Jerez, M.; Casal, S.; Sineiro, J.; Nunez, M.J.; Oliveira, M.B.P.P. Antiradical activity, phenolics profile, and hydroxymethylfurfural in espresso coffee: Influence of technological factors. *J. Agric. Food Chem.* **2010**, *58*, 12221–12229. [CrossRef] [PubMed]

30. Costa, A.S.G.; Alves, R.C.; Vinha, A.F.; Barreira, S.V.P.; Nunes, M.A.; Cunha, L.M.; Oliveira, M.B.P.P. Optimization of antioxidants extraction from coffee silverskin, a roasting by-product, having in view a sustainable process. *Ind. Crops Prod.* **2014**, *53*, 350–357. [CrossRef]

31. Barros, L.; Baptista, P.; Ferreira, I.C.F.R. Effect of *Lactarius piperatus* fruiting body maturity stage on antioxidant activity measured by several biochemical assays. *Food Chem. Toxicol.* **2007**, *45*, 1731–1737. [CrossRef] [PubMed]

32. Benzie, I.F.; Strain, J.J. The ferric reducing ability of plasma (FRAP) as a measure of "antioxidant power": The FRAP assay. *Anal. Biochem.* **1996**, *239*, 70–76. [CrossRef] [PubMed]

33. Akhtar, N.; Shoaib Khan, H.M.; Iqbal, A.; Khan, B.A.; Bashir, S. *Glycyrrhiza glabra* extract cream: Effects on skin pigment "melanin". In Proceedings of the International Conference on Bioscience, Biochemistry and Bioinformatics, Singapore, 26–28 February 2011; Volume 5, pp. 434–439.

34. Jones, D.S.; Muldoon, B.C.; Woolfson, A.D.; Sanderson, F.D. An examination of the rheological and mucoadhesive properties of poly(acrylic acid) organogels designed as platforms for local drug delivery to the oral cavity. *J. Pharm. Sci.* **2007**, *96*, 2632–2646. [CrossRef] [PubMed]

35. Rozman, B.; Gasperlin, M. Stability of vitamins C and E in topical microemulsions for combined antioxidant therapy. *Drug Deliv.* **2007**, *14*, 235–245. [CrossRef] [PubMed]

36. Wong, S.P.; Leong, L.P.; William Koh, J.H. Antioxidant activities of aqueous extracts of selected plants. *Food Chem.* **2006**, *99*, 775–783. [CrossRef]

37. *ISO Standard 11930:2012. Cosmetics—Microbiology—Evaluation of the Antimicrobial Protection of a Cosmetic Product*; International Organization for Standardization (ISO): Geneva, Switzerland, 2012.

38. Anchisi, C.; Maccioni, A.M.; Sinico, C.; Valenti, D. Stability studies of new cosmetic formulations with vegetable extracts as functional agents. *Farmaco* **2001**, *56*, 427–431. [CrossRef]

39. Bouftira, I.; Abdelly, C.; Sfar, S. Characterization of cosmetic cream with *Mesembryanthemum crystallinum* plant extract: Influence of formulation composition on physical stability and anti-oxidant activity. *Int. J. Cosmet. Sci.* **2008**, *30*, 443–452. [CrossRef] [PubMed]

40. Guaratini, T.; Gianeti, M.D.; Campos, P.M. Stability of cosmetic formulations containing esters of vitamins E and A: Chemical and physical aspects. *Int. J. Pharm.* **2006**, *327*, 12–16. [CrossRef] [PubMed]

Review

Macroalgae-Derived Ingredients for Cosmetic Industry—An Update

Filipa B. Pimentel *, Rita C. Alves *, Francisca Rodrigues and M. Beatriz P. P. Oliveira

REQUIMTE/LAQV, Department of Chemical Sciences, Faculty of Pharmacy, University of Porto, Rua de Jorge Viterbo Ferreira nr. 228, 4050-313 Porto, Portugal; franciscapintolisboa@gmail.com (F.R.); beatoliv@ff.up.pt (M.B.P.P.O.)
* Correspondence: filipabpimentel@gmail.com (F.B.P.); rita.c.alves@gmail.com (R.C.A.);
    Tel.: +351-220-428-640 (F.B.P. & R.C.A.)

Received: 31 October 2017; Accepted: 8 December 2017; Published: 25 December 2017

Abstract: Aging is a natural and progressive declining physiological process that is influenced by multifactorial aspects and affects individuals' health in very different ways. The skin is one of the major organs in which aging is more evident, as it progressively loses some of its natural functions. With the new societal paradigms regarding youth and beauty have emerged new concerns about appearance, encouraging millions of consumers to use cosmetic/personal care products as part of their daily routine. Hence, cosmetics have become a global and highly competitive market in a constant state of evolution. This industry is highly committed to finding natural sources of functional/bioactive-rich compounds, preferably from sustainable and cheap raw materials, to deliver innovative products and solutions that meet consumers' expectations. Macroalgae are an excellent example of a natural resource that can fit these requirements. The incorporation of macroalgae-derived ingredients in cosmetics has been growing, as more and more scientific evidence reports their skin health-promoting effects. This review provides an overview on the possible applications of macroalgae as active ingredients for the cosmetic field, highlighting the main compounds responsible for their bioactivity on skin.

Keywords: macroalgae; skin care; technological ingredients; antiaging; antioxidant; whitening; moisturizing; collagen boosting; photoprotection; anti-inflammatory

1. Introduction

The world population continues to grow, although at a slower rate than in the recent past, and is expected to reach 9.7 billion by 2050. Globally, demographic projections indicate that life expectancy at birth is increasing [1], which means that populations are getting older [2]. This will certainly have wide-ranging repercussions on social, economic, and health systems.

Aging is a natural and progressive declining physiological process, influenced by multifactorial aspects, that affects individuals' health in very different ways [3,4]. Oxidative stress has a substantial role in aging, and several studies have suggested different mechanisms by which free radicals can damage biological systems, leading to the development of chronic diseases: diabetes, cognitive decline and neurodegenerative diseases (e.g., Alzheimer's and Parkinson's), cardiovascular injuries, skin damage, and certain types of cancer, among many others [2,4–9].

1.1. Cosmetics Industry

The skin has historically been used for the topical delivery of compounds, being a dynamic, complex, integrated arrangement of cells, tissues, and matrix elements that regulates body heat and water loss, whilst preventing the invasion of toxic substances and microorganisms. Structurally, skin is composed of three major regions: epidermis, dermis, and hypodermis. According to Mathes et al. [10], the most superficial layer of the epidermis (*stratum corneum*) contains a cornified layer of protein-rich dead cells embedded in a lipid matrix which, in turn, manly comprises ceramides, cholesterol, and fatty acids (FA). Within the epidermis, melanocytes, Merkel cells, and Langerhans cells can also be found; these are responsible for melanin production, sensorial perception, and immunological defense, respectively [11]. The viable epidermis (50–100 μm) containing the basal membrane presents laminins (at least one type), type IV collagen, and nidogen, as well as the proteoglycan perlecan, while in the dermis (1–2 mm) it is possible to find sweat glands and hair follicles [12]. The dermis is pervaded by blood and lymph vessels. This skin layer matrix comprises not only arranged collagen fibers and a reticular layer with dense collagen fibers arranged in parallel to the skin surface, but also collagen and elastin, which provide the elastic properties of the skin [10]. Fibroblasts are the main cell type of the dermis. Beneath the dermis lies the hypodermis, where adipocytes are the most prominent cell type.

The efficacy of cosmetic active ingredients is related to their diffusion rate through the skin barrier to their specific targets [13]. However, small soluble molecules with simultaneous lipophilic and hydrophilic properties have a greater ability to cross the *stratum corneum* than do high-molecular-weight particles, polymers, or highly lipophilic substances [14]. Also, it should be highlighted that the skin surface has long been recognized to be acidic, with a pH of 4.2–5.6, being described as the acid mantle [15].

During aging, skin becomes thinner, fragile, and progressively loses its natural elasticity and ability to maintain hydration [16], and with the new society paradigms regarding youth and beauty have emerged new concerns about appearance. The use of cosmetic/personal care products (PCP) and their ingredients is part of the daily routine of millions of consumers. PCP can be locally applied on the skin, lips, eyes, oral cavity, or mucosa, but systemic exposure to the ingredients cannot be neglected and should be carefully considered. Besides this, there is the possibility of local adverse reactions, such as irritation, sensitization, or photoreactions. Given the massive use of these products, they must be diligently evaluated for safety prior to marketing [17].

According to Regulation European Commission (EC,) 1223/2009, a cosmetic product is defined as "any substance or mixture intended to be placed in contact with the external parts of the human body (epidermis, hair system, nails, lips and external genital organs) or with the teeth and the mucous membranes of the oral cavity with a view exclusively or mainly to cleaning them, perfuming them, changing their appearance, protecting them, keeping them in good condition or correcting body odours" [18].

However, there is another category—the "cosmeceuticals"—which is attracting the industry's attention and is of interest to the most attentive consumers. The term has its origin about three decades ago [19], but to this day it has no legal meaning, namely, under the Federal Food, Drug, and Cosmetic Act [20]. Even so, the industry continues to use this designation, and cosmeceuticals' development and marketing still lies between the individual benefits of cosmetics and pharmaceuticals [21]. Recently, Kim [22] stated that some cosmetic formulations, in fact, are intended to prevent disease or to affect human skin's function or structure, and can be considered as drugs. These may include sunscreens or antidandruff shampoos, but also other cosmetics containing active ingredients that promote physiological changes in skin cells, making them appear healthier and younger [23].

Cosmetics are a global and highly competitive market worth more than €425 billion worldwide [24]. In 2016, the European cosmetics and personal care market was the largest in the world, valued at €77 billion in retail sales price, followed by the United States (€64 billion) and Brazil (€24 billion) [25].

In recent decades, consumers have been drawing more and more attention not only to lifestyle issues and their impact on health and well-being, but also to environment and sustainability matters, questioning the origin of products, manufacturing processes, and ecological implications, along with safety issues [23,26]. The search for natural products for a great diversity of purposes, including food, nutraceuticals, cosmetics, and personal hygiene products, among others, somehow reflects these concerns [27]. In part, this is due to the consumers' perception about the safety of botanicals, which are derived from nature, making them desirable ingredients over synthetic ones for a diversity of formulations [16,28]. This is strong encouragement for industry-related research to find solutions and novel/alternative natural raw materials with additional properties that go further than their basic functions (e.g., nutrition) [28–30]. Nevertheless, it is of huge importance to guarantee that the selected raw materials are nontoxic and safe, and to also ensure accurate controls throughout all the production phases of industrial batches [16].

At the end, the main challenge of this whole process is to add value to products. This can be accomplished in several different ways, namely, by (i) finding natural raw materials that are simultaneously rich in functional and bioactive compounds; (ii) using these resources in a sustainable way; (iii) processing them through green processes and eco-friendly procedures, with low environmental impact; and/or (iv) delivering products and innovative solutions that meet consumers' expectations.

The following sections will provide an overview about the possible application of macroalgae as active ingredients in the cosmetic field, highlighting the main compounds responsible for their bioactivity on skin.

1.2. Macroalgae in the World—Global Numbers

Macroalgae are an excellent example of a natural resource that can fit the above-mentioned principles. According to the latest available statistics from FAO (Food and Agriculture Organization), about 23.8 million tons of macroalgae ($6.4 billion) and other algae are harvested annually. The major producing countries are China (54%) and Indonesia (27%), followed by the Philippines (7.4%), Republic of Korea (4.3%), Japan (1.85%), and Malaysia (1.39%) [31]. In Asian countries, macroalgae are traditionally used as food, for medicinal purposes, or as fertilizers. Besides this, they are a valuable raw material used as an ingredient in animal feed [31–34]. However, some authors consider them to be still underexploited and to have not yet reached their full potential of application [35].

Overall, following global trends, there is a growing demand for edible algae and algae-based products [36]. With aquaculture, which is one of the fastest growing producing sectors, it is possible to considerably increase the availability of that biomass [30,31,37].

The marine environment is extremely demanding, competitive, and aggressive. Consequently, marine organisms, including macroalgae, are forced to develop an efficient metabolic response as a self-defense mechanism, for example, by producing secondary metabolites that allow them to preserve their survival and protect themselves against external threats [32]. Therefore, sea biodiversity presents the opportunity to explore these molecules and find novel and natural bioactive compounds.

Macroalgae are one of the most ecologically and economically important living resources of the oceans, being generally classified into three groups according to their pigmentation: Phaeophyceae (brown), Rhodophyceae (red), and Chlorophyceae (green) [33,34]. Undeniably, they have huge potential as a natural source of important nutrients, namely, fiber (15–76% dry weight, dw), protein (1–50% dw), essential amino acids, essential minerals, and trace elements (ash: 11–55% dw) [24]. Despite having a low fat content (0.3–5% dw), they provide long-chain polyunsaturated fatty acids from the n-3 family (n-3 LC-PUFA), such as eicosapentaenoic acid (EPA, 20:5n-3), and liposoluble vitamins (e.g., β-carotene, vitamin E) [34,36,38–40]. However, it is important to highlight that macroalgae development and composition is affected by the species genetics and the surrounding growth conditions, namely, light, temperature, pH, salinity, and nutrient variations [35,38,41,42].

The production of macroalgae in aquaculture is not very complex and can be performed at a large scale. They can develop quickly and, by controlling their growth conditions, it is possible to manipulate their chemical composition, namely, protein, polyphenol, and pigment contents [16,41].

Regardless of their origin (either from wild harvest or from controlled production), the overall chemical composition of macroalgae makes them a very worthy bio-sustainable ingredient for a wide range of applications. This is of particular interest for the cosmetic industry, in which the ingredients used in the formulations—either active substances, excipients, or additives—are elements of added value and differentiation of a final product. The active ingredient is responsible for the cosmetic activity of interest (moisturizing, whitening, antiaging, etc.), while the excipient constitutes the vector for the active ingredient and, in turn, the additive is an ingredient intended to improve the product preservation or its organoleptic properties [24].

2. Macroalgae as a Source of Functional and Technological Ingredients

For years, due to their composition, some species of macroalgae have been traditionally used as a source of phycocolloids, namely, agar and carrageenan extracted from red algae such as *Gracilaria*, *Chondrus*, *Gelidiella*, among others, and alginate from brown algae like *Ascophyllum*, *Laminaria*, or *Sargassum* [30,33,43,44]. These phycocolloids are water-soluble polysaccharides, mainly used to thicken (increase the viscosity of) aqueous solutions, to make gels of variable degrees of firmness, to produce water-soluble films, and to stabilize some products [43]. Agar and carrageenan form thermally reversible gels, while alginate gels do not melt on heating. These compounds are industrially extracted and, due to their technological characteristics, further used as ingredients/additives in a wide variety of products in agro-food, pharmaceutic and cosmetic industries [30,31,33,43]. Table 1 describes some examples of industrial applications and technological functions of the above-mentioned phycocolloids extracted from macroalgae, specifically in the cosmetic industry [45–47].

Natural plant extracts can be incorporated in a wide variety of cosmetic products, like creams and body lotions, soaps, shampoos, hair conditioners, toothpastes, deodorants, shaving creams, perfumes, and make-up, among others; this has been a very active area of research [28,48]. Regarding, specifically, the use of macroalgae, some species are suitable for dermocosmetic applications [49].

Within the additives class, preservatives are one of the most representative substances. For the industry, finding sources of natural additives as alternatives to current commercial synthetic ones is a matter of great interest [50]. Some of the more commonly used additives are BHT (butyl hydroxytoluene) and BHA (butyl hydroxyanisole), used as synthetic antioxidants to retard lipid oxidation [51]. However, BHT has been associated with cancer and respiratory and behavioral issues in children. An alternative is to use BHA instead, although, in high doses, it can also be carcinogenic [52]. Alternatively, natural antioxidants from plants and macroalgae have been demonstrating a solid substitution potential [35,53]. Their antioxidant-rich extracts actively protect formulations against oil oxidative processes, particularly those containing a higher amount of oily phase, while simultaneously presenting health-promoting effects [27].

Currently, the interest of the cosmetic industry in macroalgae goes further than just using it as a source of excipients and additives, as those previously mentioned in Table 1. Besides their functional and technological properties, macroalgae are a source of bioactive compounds of added value, which can also be a competitive advantage for this industry.

3. Macroalgae as a Source of Bioactive Skin Care Compounds

Over the years, many studies have been conducted about the nutritional composition, secondary metabolites and bioactivities—as well as the potential health-promoting effects—of macroalgae. To date, most of these marine-derived compounds were intended for food and pharmaceutical applications [38]. Also, several researchers have been exploring the effects of macroalgae on health, showing some progress and important positive outcomes in regards to some types of cancer; heart diseases; thyroid and immune functions; allergy; inflammation [54]; and antioxidant, antibacterial, and antiviral activity [55], among many others [42,56,57].

Aware of this, the cosmetic industry is interested in using macroalgae as a source of bio-sustainable ingredients since they are extremely rich in biologically active compounds (Table 2), some of which are already documented as functional active skin care agents [24,58]. As an additional advantage for this industry, these ingredients can be cheap, while matching consumers' requests for "natural" and "healthier" products.

Cosmetics **2018**, *5*, 2

Table 1. Applications and technological functions of phycocolloids extracted from macroalgae in the cosmetics industry.

Ingredient	Species	Technological Function	Application	Reference
Carrageenan	Not specified	Thickening and gelling agent, binder, sensory enhancer	Bath and shower gel	[45]
Carrageenan	Not specified	Thickening and suspending agent, stabilizer, sensory enhancer	Skin care	[45]
Carrageenan	Not specified	Thickening and suspending agent, stabilizer, sensory enhancer	Sun care	[45]
Carrageenan	Not specified	Thickening agent, film former, fixative agent, sensory enhancer	Hair care	[45]
Alginate	Not specified	Interface vitalization	Shampoo	[45]
Carrageenan	Not specified	Thickening and suspending agent, stabilizer, binder	Oral care	[45]
Alginate	Not specified	Form retention	Dental moulds	[45]
Alginate	Not specified	Emulsification, viscosity	Lipstick	[45]
Gelcarin® PC 379	*Chondrus crispus*	Exfoliant	Decorative cosmetic care applications	[46]
Gelcarin® PC 812	*Chondrus crispus*	Emulsifier and thickener	Lipsticks and deodorants	[46]
Wakamine 1% (peptidic extract)	*Undaria pinnatifida*	Whitening/lightening agent	Skin care products	[46]
Wakamine XP	*Undaria pinnatifida*	Whitening/lightening agent	Skin care products	[46]
EPHEMER™	*Undaria pinnatifida*	Antioxidant and anti-aging agent	Skin care products	[46]
Akomarine® Fucus	*Fucus vesiculosus*	Skin softness and elasticity	Slimming and anti-cellulitis cosmetic formulations	[46]
DENTACTIVE®	*Fucus serratus*	Protecting agent (reduces gingivorrhagia)	Oral-care products	[46]
Gracilaria Hydrogel	*Gracilaria conferta*	Humectant, nourishing and conditioning agent	Skin care products	[46]
Hijiki Extract	*Hizikia Fusiforme*	Whitening agent	Whitening preparations	[46]
Chlorofiltrat® Ulva HG	*Ulva lactuca*	Moisturizing and anti-inflammatory agent	Skin care products	[46]
AT UV PROTECTOR P	*Porphyra tenera*	Photo-protection	Skin and sun care	[46]
XYLISHINE™	*Pelvetia canaliculata*	Hair moisturizer	Hair formulations	[47]

Table 2. Health benefits associated with macroalgae-derived bioactive compounds.

Bioactive Compounds	Species	Assays	Health Benefits	Reference
Polysaccharides				
Sulphated oligosaccharides or polysaccharides	*Solieria chordalis* [1]	*In vitro* assays against the *Herpes simplex* virus in African green monkey kidney cells (Vero, ATCC CCL-81); cell viability study by neutral red assay, using Vero cell/HSV-1 model	Some fractions obtained from *Solieria chordalis* with good antiherpetic activities; no cytotoxicity observed	[59]
Sulphated polysaccharides (carrageenans)	*Gigartina acicularis* [1] *Gigartina pistillata* [1] *Eucheuma cottonii* [1] *Euchema spinosa* [1]	Antioxidant activity assays—superoxide anion and hydroxyl radical scavenge capacity, and *in vitro* study in liver microsomal lipid peroxidation	High antioxidant activity and free radical scavenging activity, especially shown by lambda carrageenan	[60]
Sulphated polysaccharides	*Pterocladia capillacea* [1]	*In vitro* assays for antioxidant capacity and antibacterial effect against *Escherichia coli* and *Staphylococcus aureus*	Antioxidant and antibacterial activity	[61]
Sulphated polysaccharides	*Porphyra haitanensi* [1] *Laminaria japonica* [2] *Ulva pertusa* [3] *Enteromorpha linza* [3] *Bryopsis plumose* [3]	Antioxidant activity assays—superoxide and hydroxyl radical scavenging effects, and reducing power	Antioxidant response is dependent on the type of polysaccharides, which differs among red, brown, and green species	[62]
Not specified	*Ecklonia cava* [2]	*In vitro* study in murine colon cancer cell line (CT-26), mouse melanoma cell line (B-16), hamster fibroblast cell line/normal cell line (V79-4) and human leukaemia cell lines (U-937 and THP-1)	Strong selective cell proliferation inhibition on all cancer cell lines tested, high antioxidant activity, and low cell toxicity	[51]
Sulphated polysaccharides: homofucans	*Fucus vesiculosus* [2] *Padina gymnospora* [2]	Antioxidant activity assays—superoxide anion and hydroxyl radical scavenge capacity, and *in vitro* study in liver microsomal lipid peroxidation	High antioxidant activity and free radical scavenging activity	[60]
Not specified	*Rhizoclonium hieroglyphicum* [3]	Moisturizing effect in pig skin model and in human skin (human volunteers)	Increased moisturizing effect, comparable to hyaluronic acid; no skin irritation observed	[63]
Not specified	*Laminaria japonica* [2]	*In vivo* skin moisturizing activity	Increased moisturizing skin effect	[64]
Sulphated oligosaccharides or polysaccharides	*Ulva sp.* [3]	*In vitro* assays against the *Herpes simplex* virus in African green monkey kidney cells (Vero, ATCC CCL-81); cell viability study by neutral red assay, using Vero cell/HSV-1 model	Some fractions obtained from *Solieria chordalis* with good antiherpetic activities; no cytotoxicity observed	[59]

Legend: [1], Rhodophyta; [2], Ochrophyta; [3], Chlorophyta.

Some of the bioactive compounds associated with skin care include polysaccharides, proteins, (especially peptides and amino acids), lipids (including PUFA, sterols, and squalene), minerals, and vitamins, but also the secondary metabolites such as phenolic compounds, terpenoids, and halogenated compounds, among others [22,24,32,42,65]. Depending on their physicochemical properties, molecular size, and solubility, bioactive compounds can be extracted, isolated, and purified by several different methods [65]. However, in order to be used as ingredients in cosmetics, solvents used in the whole process of extraction must be GRAS-grade (Generally Recognized As Safe), which excludes all of those listed as substances prohibited in cosmetic products, described in Annex II of Regulation (EC) No 1223/2009 of the European Parliament and of the Council of 18 December 2006, concerning cosmetic products [18]. Table 2 summarizes some health benefits associated with macroalgae-derived bioactive compounds.

3.1. Polysaccharides

The biological activity of several macroalgae-derived sulphated polysaccharides (SPs) has been often reported [65,66]. The chemical structure of these macromolecules varies according to the species: brown species present mainly laminarans (up to 32%–35% dw) and fucoidans; red algae are mainly rich in carrageenans and porphyrans; and green algae are typically rich in ulvans [50,55,67]. Anti-proliferative activity in cancer cell lines as well as inhibitory activity against tumors has been described for fucoidans [66]. The genus *Porphyra* contains mainly porphyrans, an agar-like sulphated galactan disaccharide, which accounts for up to 48% of thallus (dw) [35]. It has been reported that red macroalgae SPs, namely, xylomannan, galactans, and carrageenans, exhibit antiviral activity [59]. When accessing the antioxidant activity of different SPs—carrageenans (lambda, kappa, and iota), fucoidans, and fucans—de Souza and colleagues [60] found that fucoidan and lambda carrageenan exhibited the highest antioxidant activity and free radical scavenging activity against superoxide anions and hydroxyl radicals. Ulvans, in turn, designated a water-soluble group typically found in green macroalgae, which are mainly composed of glucuronic acid and iduronic acid units together with rhamnose and xylose sulphates [55]. It has been reported that these compounds present a high antioxidant capacity against some reactive oxygen species (ROS), namely, superoxide and hydroxyl radicals [62].

3.2. Proteins, Peptides, and Amino Acids

In macroalgae, proteins are a structural part of cell walls, enzymes, and bioactive molecules, such as glycoproteins and pigments [68].

3.2.1. Protein

Protein content is one important parameter when determining the value of biomass, and may be the starting point for selecting species that may be more profitable from which to obtain bioactive peptides and amino acids through selected enzymes. The interest in enzymes in the field of cosmetics has increased. Enzymes are highly specific and selective, easy to process, and can be applied in a wide range of substrates and organic transformations in diverse reaction media [69].

Besides presenting substantial amounts of protein (up to 47% dw), most species present a complete profile of essential amino acids [70]. Even so, protein content varies according to species, being generally higher in Rhodophyceae (8%–50% dw), compared with Chlorophyceae and Phaeophyceae (7%–32% and 6%–24% dw, respectively) [29]. Geographical origin and seasonality also affect their protein composition, especially because nitrogen availability may fluctuate due to water temperature and salinity variations, light irradiation, and wave force, thereby affecting their nutrient supply [29,71].

3.2.2. Peptides

Peptides are formed of short chains of 2 to 20 amino acids. Their biofunctional properties depend on their amino acid composition and sequence in the parent protein, which needs to undergo a hydrolysis, commonly with digestive enzymes, so that peptides can be released and become active [29]. The biofunctional and bioactive properties of peptides are based on their physiological behavior, which resembles hormones or druglike activities. Besides this, they have the capacity to modify physiological functions, even in the skin, due to their ability to interact with target cells, binding to specific cell receptors or inhibiting enzymatic reactions [72,73]. Marine peptides, including macroalgae-derived ones, have been considered safer than synthetic molecules due to their high bioactivity and biospecificity to targets, with rare adverse effects and reduced risk of unwanted side effects [74]. In fact, lately, peptides have been considered a captivating topic in the field of cosmetics and skin applications [75].

3.2.3. Amino Acids

Macroalgae are an excellent source of amino acids and amino acid derivatives, which constitute the natural moisturizing factor (NMF) in the *stratum corneum* and promote collagen production in the skin [65]. Some species of red macroalgae like *Palmaria* and *Porphyra* have been reported to present high amounts of arginine in their composition. Arginine is a precursor of urea, which is a component of NMF, used in cosmetic formulations [65].

Mycosporine-like amino acids are a family of secondary water-soluble metabolites with low molecular weight [29]. They are characterized by a cyclohexenone or cyclohexenimine chromophore conjugated with a nitrogen substituent of an amino acid, amino alcohol, or amino group, with maximum absorption wavelengths ranging from 310 to 360 nm [76]. Mycosporine-like amino acids protect macroalgae from UV radiation, and have been described as important antioxidant compounds in red algae with reports that they are very efficient photoprotector agents [29,65,77]. Hence, these metabolites have great potential to be used as natural skin protection ingredients in photo-protective formulations.

3.3. Lipids

Macroalgae are known as a low-energy food, and, although their total lipid contribution is generally very low (<4.5% dw), their PUFA contents are comparable to or even higher than those found in terrestrial plants [38,78]. Still, the main classes of lipids are present in their composition and include essential FA, triglycerides, phospholipids, glycolipids, sterols, liposoluble vitamins (A, D, E, and K), and carotenoids [65].

Long-chain PUFAs (LC-PUFAs) have 20 or more carbons with two or more double bonds from the methyl (omega) terminus. Marine lipids contain substantial amounts of LC-PUFAs, among which eicosapentaenoic acid (EPA; 20:5n-3) and docosahexaenoic acid (DHA; 22:6n-3) are the most important (Figure 1), along with the precursors α-linolenic acid (ALA; 18:3n-3) and docosapentaenoic acid (22:5n-3) [38]. Beneficial clinical and nutraceutical applications have been described for these compounds [78].

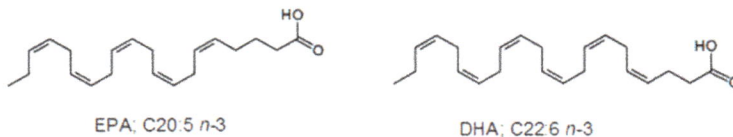

EPA: C20:5 n-3   DHA: C22:6 n-3

Figure 1. Chemical structure of eicosapentaenoic acid (EPA; C20:5n-3) and docosahexaenoic acid (DHA; C22:6n-3).

LC-PUFAs are essential components of all cell membranes and eicosanoid precursors, and are critical bioregulators of many cellular processes [79]. As mediators of many different biochemical pathways, they play an important role in health [80]. In several macroalgae species, EPA (C20:5n-3) is frequently the most representative PUFA—in some cases, reaching 50% of the total FA content [78].

A study performed by Kumari and colleagues [78] reported interesting features when comparing several macroalgae species: Chlorophyta species presented higher C18-PUFAs amounts than did C20-PUFAs, while the analyzed Rhodophyta species showed the opposite trend. In turn, Phaeophyta samples exhibited a C18-PUFAs profile comparable to that of Chlorophyta and a C20-PUFAs profile similar to that of Rhodophyta. Both brown and red species were richer in arachidonic acid and EPA, while the green ones contained higher amounts of DHA.

As stated by several authors, variations in the lipid content and FA composition are often found, and it is generally accepted that such disparities, besides the already mentioned environmental conditions, could be due to different sample treatments and extraction methods [81,82].

3.4. Vitamins and Minerals

Macroalgae are a good source of both fat-soluble vitamins (e.g., vitamin E) and water-soluble vitamins, namely, B1 (thiamine), B2 (riboflavin), B3 (niacin), B5 (pantothenic acid), B6 (pyridoxine), B12 (cobalamin), B8 (biotin), B9 (folic acid), and C (ascorbic acid) [40]. Interestingly, macroalgae are also one of the few vegetable sources of vitamin B12—Its presence is likely due to the bacteria living on their surface or in the proximate waters [35].

Besides this, macroalgae are important sources of minerals and trace elements, namely, calcium, sodium, potassium, magnesium, iron, copper, iodine, and zinc [40].

3.5. Pigments and Phenolic Compounds

Macroalgae contain a wide variety of pigments that absorb light for photosynthesis, many of which are not found in terrestrial plants. Species are characterized by specific sets of pigments. Three major classes of photosynthetic pigments are found in algae: chlorophylls, carotenoids (carotenes and xanthophylls), and phycobiliproteins [68]. These compounds are responsible for macroalgae color variations during their growth and reproduction cycles, which depend on the amounts of pigments present (chlorophylls, carotenoids, and their breakdown metabolites) [83].

Chlorophylls and carotenoids are liposoluble molecules. Chlorophylls, the greenish pigments, are a group of cyclic tetrapyrrolic pigments, with a porphyrin ring with a central magnesium ion and usually a long hydrophobic chain. Generally, chlorophyll a is the most abundant photosynthetic pigment, while others are considered accessory pigments [68].

In turn, carotenoids are polyene hydrocarbons biosynthesized from eight isoprene units (tetraterpenes) [84], usually presenting red, orange, or yellow colorations and remarkable antioxidant properties [68]. Within the carotenes group, β-carotene is the most representative one and is present in all classes of macroalgae [68].

Xanthophylls contain oxygen in the form of hydroxy, epoxy, or oxo groups [84]. The foremost representative marine xanthophylls include astaxanthin and fucoxanthin (Figure 2), which have also been recognized to have excellent antioxidant potential [50].

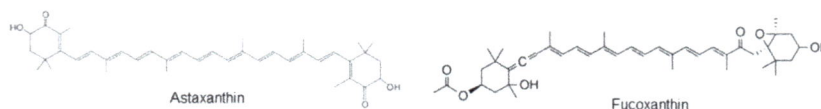

Astaxanthin

Fucoxanthin

Figure 2. Chemical structure of the most representative marine xanthophylls—astaxanthin and fucoxanthin.

Astaxanthin is a lipophilic carotenoid, structurally similar to β-carotene but with an additional hydroxyl and ketone group on each ionone ring [50]. Some studies have reported that astaxanthin can be more effective than β-carotene in preventing lipid peroxidation in solution and various biomembrane systems [85]. In turn, fucoxanthin is one of the major xanthophyll pigments in brown algae and is found in edible brown algae, such as *Undaria* sp., *Sargassum* sp., *Laminaria* sp., and

Hizikia sp. [35]. This molecule presents a unique structure including allenic, conjugated carbonyl, epoxide, and acetyl groups, and was recently identified as the major bioactive antioxidant carotenoid in 30 Hawaiian macroalgae species [86].

Phycobiliproteins (PBP) are a water-soluble group of photosynthetic pigments comprising different compounds, like phycoerythrins with a red pigment linked to the protein molecule, or phycocianins with a blue pigment instead. These different molecules absorb at different wavelengths of the spectrum, which makes them very colorful and highly fluorescent *in vivo* and *in vitro*. This is of special interest for biotechnological applications, where they are useful in diverse biomedical diagnostic systems (e.g., immunochemical methods) [87]. Some have been used as natural food colorants, as well as pink and purple dyes in lipsticks, eyeliner, and other cosmetic formulations [65,88,89]. Being water-soluble molecules, it is possible to extract PBP from algal tissues using green extraction solvents, like water or buffers [68].

In macroalgae, phenolic compounds are secondary metabolites, which means that they do not directly intervene in primary metabolic processes such as photosynthesis, cell division, or reproduction [65]. Instead, it is believed that this class of compounds is mainly responsible for protection mechanisms, namely, against oxidative stress or UV cytotoxic effects [35,65,68].

Phlorotannins, a subgroup of tannins mainly found in brown macroalgae and, to a lesser extent, in red species, are derived from phloroglucinol units (1,3,5-trihydroxybenzene), whereas in plants polyphenols are derived from gallic and ellagic acids [22,35]. Phlorotannins are highly hydrophilic compounds with a wide range of molecular sizes (from 126 Da to 650 kDa), and are of interest for different applications (e.g., nutritional supplements, cosmetic and cosmeceutical products) [22].

4. Skin Benefits

Many external factors, including UV radiation, climate conditions, and air/environmental pollutants (e.g., tobacco smoke) can affect the protective ability of skin and promote its premature aging [90]. Commonly, this continuous exposure leads to oxidative stress caused by the imbalance between oxidants and antioxidants, which affects skin health [90]. Skin aging produces several changes: it becomes thinner, more fragile, and progressively loses its natural elasticity and ability to maintain hydration [16].

In cosmetic formulations, the primary functions of natural ingredients may be antioxidant, collagen boosting, or even anti-inflammatory [91]. The incorporation of macroalgae-derived bioactive compounds in cosmetics has been growing as more and more scientific evidence is documented in regards to their health-promoting and anti-pollution effects [55,62,82]. The foremost interesting classes of bioactive compounds include those intended for antiaging care, including protection against free radicals, prevention of skin flaccidity and wrinkles, anti-photoaging, photoprotection against UV radiation, moisturizing, and skin whitening [16,27,65].

4.1. Antiaging and Antioxidant Effects

In biological systems, oxygen is the most common generator of free radicals—highly reactive molecules with harmful potential. ROS and reactive nitrogen species (RNS, such as nitric oxide, $NO^{\bullet}$) are products of normal cellular metabolism. They act as secondary messengers by regulating several normal physiological functions. However, they can play a dual role, as they can act as both damaging and beneficial species. Oxidative stress, caused by an overproduction of ROS, can induce serious damages in several cell structures (lipids and membranes, proteins, and DNA). At the same time, ROS and RNS also participate in several redox regulatory mechanisms of cells in order to protect them against oxidative stress and maintain their "redox homeostasis" [9].

A great diversity of bioactive compounds, namely vitamin E, vitamin C, superoxide dismutase, coenzyme Q10, zinc sulphate, ferulic acid, polyphenols, and carotenoids, among others, have been successfully used, for a long time, in cosmetic products as free-radical-scavenging molecules [23].

An *in vitro* study showed that an algal extract containing astaxanthin presented a protective effect in the reduction of DNA damage and maintenance of cellular antioxidant status in lines of human skin fibroblasts (1BR-3), human melanocytes (HEMAc), and human intestinal Caco-2 cells, irradiated with UVA [85].

In the last few years, other classes of macroalgae compounds have been showing potential as bioactive ingredients for cosmetics. In a study performed with *Ecklonia cava*, crude polysaccharide and polyphenolic fractions obtained by a former enzymatic hydrolysis were evaluated, showing a suppressive effect on tumor cell growth, and antioxidant and radical scavenging activities in different cell lines, with low toxicity [51]. In another study, Zhang and colleagues [62] evaluated the antioxidant activity of SPs extracted from five macroalgae—one brown (*Laminaria japonica*), one red (*Porphyra haitanensis*), and three green species (*Ulva pertusa*, *Enteromorpha linza* and *Bryopsis plumose*)—reporting that their antioxidant behavior depended on the type of polysaccharides of each extract, which was shown to be different among the species [62].

Likewise, protein hydrolysates, peptides, or amino acids from macroalgae can play a substantial antioxidant role in a diverse range of oxidative processes [72].

4.2. Moisturizing/Hydration Action

Moisturizing and hydration are crucial for skin care and are essential to maintaining its healthy appearance and elasticity, while also strengthening its role as a barrier to harmful environmental factors [65]. Approximately 60% of the epidermis is water which is fixed by hygroscopic substances known by the generic name of NMF (natural moisturizing factor). NMF constitution includes amino acids (40%), including serine (20–30%), lactic acid (12%), pyrrolidone carboxylic acid (12%), urea (8%), sugars, minerals, and a fraction that still remains undetermined [24]. Topical application of the above-mentioned components, which can act as humectants, can improve the skin moisturizing ability and relieve a dry skin condition [92].

Polysaccharides play a very important role in cosmetic formulations as humectants and moisturizers. These macromolecules have a high capacity for water storage and can be linked to keratin through hydrogen bonds, thus improving skin moisturization [63,65]. According to Wang and colleagues [93], polysaccharides extracted from *Saccharina japonica* revealed better moisturizing properties than hyaluronic acid, suggesting that these polysaccharides could be an interesting ingredient for cosmetics. The authors also found that the sulphated group was a main active site for moisture absorption and moisture retention ability, and that the lower-molecular-weight polysaccharides presented the highest moisture absorption and moisture retention abilities [93]. A cosmetic formulation containing 5–10% extract of *Laminaria japonica* improved skin moisture in a group of volunteers. Authors suggest that two mechanisms might be responsible for these promising results: on the one hand, the hydroscopic substances of the extract (e.g., free amino acids, sugars, and minerals) may contribute to reinforcing the NMF in skin, helping to retain appropriate moisture levels in the epidermis; on the other hand, phycocolloids, like alginate, and protein in extracts attach to skin proteins to form a protective barrier for moisture loss regulation [64].

4.3. Collagen-Boosting Effects

With aging, the extracellular tissue matrix (ETM) components—collagen, hyaluronic acid, and elastin, among others—decrease, leading to thinner skin with a weakened structure [94]. However, some active ingredients have been showing promising results in reverting these signs. For instance, some peptides have been used as cosmeceutical ingredients showing interesting antiaging effects, namely in wrinkle and fine line reduction, and in skin firming and skin whitening [73]. Different types of peptides and mechanisms of action are responsible for those effects. Signal peptides, for instance, stimulate ETM production by specifically increasing neocollagenesis [73]. Besides this, they can also promote fibronectin and elastin synthesis, as well as cell–cell cohesion, with results in skin firming and wrinkle and fine line reduction [73]. Therefore, the use of formulations containing these compounds can promote the replacement of the lost extracellular tissue matrix, reducing, then, the appearance of wrinkles [23].

Marine-derived phlorotannins, extracted from *Eisenia bicyclis* and *Ecklonia kurome*, presented a strong hyaluronidase inhibitory effect in *in vitro* assays [95], showing potential as a bioactive ingredient to recover ETM functions.

4.4. Photo-Protective Action

Sunlight UV radiation is still the most powerful environmental risk factor in skin cancer pathogenesis [85]. The use of photoprotective products with UV filters is extensively recommended to prevent (and protect the skin from) several types of damage, like sunburn, photo-aging, photodermatoses, or even skin cancer [27]. Within this type of product, formulations containing sun-screening agents combined with antioxidants are considered to be safer and more effective [23].

Bioactive compounds able to absorb UV radiation can protect human fibroblast cells from UV-induced cell death and suppress UV-induced *aging* in human skin [65].

As previously mentioned, macroalgae have developed mechanisms to prevent damage from UVB and UVA radiations, either by producing screen pigments, like carotenoids, or by phenolics.

4.5. UV-Absorbing Compounds

Heo and Jeon [96] reported that fucoxantin from *Sargassum siliquastrum* presented a great *in vitro* ability to protect human fibroblasts against oxidative stress induced by UVB radiation. Another study with *Halidrys siliquosa* (Phaeophyta) showed that the tested extracts presented strong antioxidant activity and a good sunscreen potential, associated with the presence of phlorotannins like diphlorethols, triphlorethols, trifuhalols, and tetrafuhalols [57].

4.6. Whitening/Melanin-Inhibiting Effects

Melanin, which is the main determinant of skin color, absorbs UV radiation and prevents free radical generation, protecting skin from sun damage and aging [97]. However, the abnormal production of melanin can be a dermatological condition and a serious cosmetic issue.

Tyrosinase catalyzes melanin synthesis in two different pathways: the hydroxylation of L-tyrosine to 3,4-dihydroxy-l-phenylalanine (L-dopa) and the oxidation of L-dopa to dopaquinone, followed by further conversion to melanin [24]. It is possible to regulate melanin biosynthesis, for instance, by protecting skin and avoiding UV exposure, or by inhibiting tyrosinase action or melanocyte metabolism and proliferation [22].

The demand for natural products that inhibit/control or prevent melanogenesis and, consequently, skin pigmentation, is growing all over the world [98], especially for melanin hyperpigmentation dermatological diseases, as well as for cosmetic formulations for depigmentation [22]. Recently, macroalgae extracts showed profound inhibitory effects against tyrosinase and melanin synthesis in both *in vitro* cell experiments [98,99] and an *in vivo* zebrafish animal model [99].

4.7. Anti-Inflammatory Effect

An inflammatory process causes oxidative stress and reduces cellular antioxidant capacity. The large amount of produced free radicals react with FA and proteins of cell membranes, permanently damaging their normal functions [4].

Senevirathne and colleagues [100] evaluated antioxidant and anticholinesterase (AChE) activities, as well as the protective effects of enzymatic extracts from *Porphyra tenera* against lipopolysaccharides (LPS)-induced nitrite production in RAW264.7 macrophage cells. The authors concluded that all enzymatic extracts showed no cell cytotoxicity (cell viabilities greater than 90% in all cases), and all enzymatic extracts effectively inhibited LPS-induced nitric oxide production in RAW264.7 macrophages [100]. These results indicate that *Porphyra tenera* could be a valuable source of natural antioxidants and anti-inflammatory ingredients for cosmetic purposes.

4.8. Anti-Cellulite and Slimming Effects

Although cellulite is not a pathological condition, it remains a matter of cosmetic concern, especially for postadolescent women [101]. Many efforts have been made to find treatments that improve symptoms and signs of cellulite, as well as the visual appearance of skin.

Some species of macroalgae (e.g., *Fucus vesiculosus* L., *Laminaria digitata* (Huds.) Lamouroux, among others) are used in cosmetic formulations for cellulite reduction [65].

Al-Bader and colleagues [102] tested a formulation containing aqueous extracts of *Furcellaria lumbricalis* and *Fucus vesiculosus* to assess *in vitro* lipolysis in mature adipocytes and measured pro-collagen I in human primary fibroblasts, finding that there was an improvement of lipolysis-related mechanisms and pro-collagen I production. Subsequently, they evaluated cellulite by dermatological grading and ultrasound measurements and could observe a clinical improvement in the cellulite.

Macroalgae extracts may also be of interest for slimming purposes, as evidence demonstrates that they significantly decrease the body weight gain, fat-pad weight, and serum and hepatic lipid levels in high-fat-diet-induced Sprague Dawley male obese rats, and showed a protective effect against these factors through the regulation of gene and protein expression involved in lipolysis and lipogenesis [103].

Iodine is essential for thyroid metabolism. Thyroid hormones are involved in mechanisms that increase the synthesis of carnitine palmitoyl transferase which, in turn, promote lipolysis by increasing the penetration of fatty acids in the mitochondria [24]. Diet is the major contributor of iodine, but breathing gaseous iodine in the air and placing it on the skin are other possible paths [104]. *Fucus serratus* L. is a rich source of iodine. A recent *in vivo* study reported that bath thalassotherapy with this macroalgae had the potential to increase the urinary iodide concentration of the bather, indicating inhalation of volatile iodine as the predominant route of uptake [104]. Another *in vivo* study also showed the effectiveness of a cosmetic product containing extracts of *Gelidium cartilagineum*, *Pelvetia canaliculata*, and *Laminaria digitata*, as well as other active ingredients, in exerting a slimming effect, compared with a placebo [105].

4.9. Antiviral and Antibacterial Effects

An enzyme-assisted extraction enabled a more effective obtention of proteins, neutral sugars, uronic acids, and sulphate groups in three species of macroalgae: the red *Solieria chordalis*, the green *Ulva* sp., and the brown *Sargassum muticum*. In this study, although no cytotoxicity was observed for all extracts, only *S. chordalis* presented good antiherpetic activities, mainly attributed to its richness in sulphate groups [59].

An O/W (oil in water) emulsion prepared with a phlorotannin-enriched fraction obtained from the brown macroalgae *Halidrys siliquosa* presented antibacterial capacity against *Pseudomonas aeruginosa*, *Staphylococcus aureus*, and *Escherichia coli* [57]. Another study with a red macroalgae (*Pterocladia capillacea*) revealed that using carbohydrate degrading enzymes prior to *in vitro* assays produced extracts with higher antioxidant capacity and stronger antibacterial effect against *Escherichia coli* and *Staphylococcus aureus* [61]. This could be of particular interest for the development of natural preservatives to be used in cosmetic formulations.

5. Final Remarks

Macroalgae-derived ingredients have been used in cosmetic formulations due to their technological properties. However, it is well established that the interest of the cosmetic industry in macroalgae goes further than just using it as a source of excipients and technological additives. Macroalgae are a source of added-value compounds, with scientific evidence showing their benefits for human health and wellbeing. This can be a competitive advantage for this industry, namely in terms of finding and using novel molecules and agents that apparently have biological effects on skin, such as antiaging, antioxidant, moisturizing, collagen-boosting, photo-protective, whitening and

Cosmetics **2018**, *5*, 2

melanin-inhibiting, anti-inflammatory, anti-cellulite and slimming, and antiviral and antibacterial activities. This review has summarized some of the possible applications of macroalgae as active ingredients in the cosmetic field, highlighting the main compounds responsible for their bioactivity on skin.

Acknowledgments: The authors thank the financial support to the project Operação NORTE-01-0145-FEDER-000011—denominada Qualidade e Segurança Alimentar- uma abordagem (nano)tecnológica. This work was also supported by the project UID/QUI/50006/2013—POCI/01/0145/FEDER/007265 with financial support from FCT/MEC through national funds and co-financed by FEDER.
F. B. Pimentel is grateful to FCT for the PhD research grant (SFRH/BD/109042/2015). Francisca Rodrigues is thankful for her post-doc research grant from the project Operação NORTE-01-0145-FEDER-000011.

Author Contributions: Filipa B. Pimentel carried out the literature research and wrote the first draft of the manuscript under the supervision of Rita C. Alves and M. Beatriz P. P. Oliveira. Francisca Rodrigues collaborated in writing a part of the manuscript. Rita C. Alves and M. Beatriz P. P. Oliveira performed the final revision of the manuscript.

Conflicts of Interest: The authors declare no conflict of interest.

References

1. Department of Economic and Social Affairs of the United Nations. *Population Division World Population Prospects: The 2015 Revision, Key Findings and Advance Tables*; Working Paper No. ESA/P/WP.241; Department of Economic and Social Affairs of the United Nations: New York, NY, USA, 2015.
2. Sieck, G.C. Physiology in perspective: Aging and underlying pathophysiology. *Physiology* **2017**, *32*, 7–8. [CrossRef] [PubMed]
3. Chatterji, S.; Byles, J.; Cutler, D.; Seeman, T.; Verdes, E. Health, functioning, and disability in older adults—Present status and future implications. *Lancet* **2015**, *385*, 563–575. [CrossRef]
4. Khansari, N.; Shakiba, Y.; Mahmoudi, M. Chronic inflammation and oxidative stress as a major cause of age-related diseases and cancer. *Recent Pat. Inflamm. Allergy Drug Discov.* **2009**, *3*, 73–80. [CrossRef] [PubMed]
5. Baierle, M.; Nascimento, S.N.; Moro, A.M.; Brucker, N.; Freitas, F.; Gauer, B.; Durgante, J.; Bordignon, S.; Zibetti, M.; Trentini, C.M.; et al. Relationship between inflammation and oxidative stress and cognitive decline in the institutionalized elderly. *Oxidative Med. Cell. Longev.* **2015**, *2015*, 804198. [CrossRef] [PubMed]
6. Brand, R.M.; Epperly, M.W.; Stottlemyer, J.M.; Skoda, E.M.; Gao, X.; Li, S.; Huq, S.; Wipf, P.; Kagan, V.E.; Greenberger, J.S. A topical mitochondria-targeted redox-cycling nitroxide mitigates oxidative stress-induced skin damage. *J. Investig. Dermatol.* **2017**, *137*, 576–586. [CrossRef] [PubMed]
7. Lin, M.T.; Beal, M.F. Mitochondrial dysfunction and oxidative stress in neurodegenerative diseases. *Nature* **2006**, *443*, 787–795. [CrossRef] [PubMed]
8. Förstermann, U.; Xia, N.; Li, H. Roles of vascular oxidative stress and nitric oxide in the pathogenesis of atherosclerosis. *Circ. Res.* **2017**, *120*, 713–735. [CrossRef] [PubMed]
9. Valko, M.; Leibfritz, D.; Moncol, J.; Cronin, M.T.D.; Mazur, M.; Telser, J. Free radicals and antioxidants in normal physiological functions and human disease. *Int. J. Biochem. Cell Biol.* **2007**, *39*, 44–84. [CrossRef] [PubMed]
10. Mathes, S.H.; Ruffner, H.; Graf-Hausner, U. The use of skin models in drug development. *Adv. Drug Deliv. Rev.* **2014**, *69*, 81–102. [CrossRef] [PubMed]
11. Prow, T.W.; Grice, J.E.; Lin, L.L.; Faye, R.; Butler, M.; Becker, W.; Wurm, E.M.; Yoong, C.; Robertson, T.A.; Soyer, H.P. Nanoparticles and microparticles for skin drug delivery. *Adv. Drug Deliv. Rev.* **2011**, *63*, 470–491. [CrossRef] [PubMed]
12. Franzen, L.; Windbergs, M. Applications of Raman spectroscopy in skin research—From skin physiology and diagnosis up to risk assessment and dermal drug delivery. *Adv. Drug Deliv. Rev.* **2015**, *89*, 91–104. [CrossRef] [PubMed]
13. Rawlings, A.; Matts, P.; Anderson, C.; Roberts, M. Skin biology, xerosis, barrier repair and measurement. *Drug Discov. Today Dis. Mech.* **2008**, *5*, e127–e136. [CrossRef]
14. Rawlings, A.V.; Matts, P.J. *Stratum corneum* moisturization at the molecular level: An update in relation to the dry skin cycle. *J. Investig. Dermatol.* **2005**, *124*, 1099–1110. [CrossRef] [PubMed]

15. Schmid-Wendtner, M.-H.; Korting, H.C. The pH of the skin surface and its impact on the barrier function. *Skin Pharmacol. Physiol.* **2006**, *19*, 296–302. [CrossRef] [PubMed]

16. Wang, H.-M.D.; Chen, C.-C.; Huynh, P.; Chang, J.-S. Exploring the potential of using algae in cosmetics. *Bioresour. Technol.* **2015**, *184*, 355–362. [CrossRef] [PubMed]

17. Nohynek, G.J.; Antignac, E.; Re, T.; Toutain, H. Safety assessment of personal care products/cosmetics and their ingredients. *Toxicol. Appl. Pharmacol.* **2010**, *243*, 239–259. [CrossRef] [PubMed]

18. Official Journal of the European Union. *European Parliament, Regulation (EC) No 1223/2009 of the European Parliament and of the Council of 30 November 2009 on Cosmetic Products*; European Union: Brussels, Belgium, 2009; pp. L-342/59–L-342/209.

19. Vermeer, B.J.; Gilchrest, B.A.; Friedel, S.L. Cosmeceuticals: A proposal for rational definition, evaluation, and regulation. *Arch. Dermatol.* **1996**, *132*, 337–340. [CrossRef] [PubMed]

20. FDA. Cosmeceutical. Available online: https://www.fda.gov/Cosmetics/Labeling/Claims/ucm127064.htm (accessed on 31 August 2017).

21. Brandt, F.S.; Cazzaniga, A.; Hann, M. *Cosmeceuticals: Current Trends and Market Analysis, Seminars in Cutaneous Medicine and Surgery, 2011*; Frontline Medical Communications: Parsippany, NJ, USA; pp. 141–143.

22. Kim, S.K. Marine cosmeceuticals. *J. Cosmet. Dermatol.* **2014**, *13*, 56–67. [CrossRef] [PubMed]

23. Zappelli, C.; Barbulova, A.; Apone, F.; Colucci, G. Effective active ingredients obtained through Biotechnology. *Cosmetics* **2016**, *3*, 39. [CrossRef]

24. Couteau, C.; Coiffard, L. Seaweed Application in Cosmetics. In *Seaweed in Health and Disease Prevention*; Fleurence, J., Levine, I., Eds.; Elsevier Inc.: Amsterdam, The Netherlands, 2016; pp. 423–441.

25. Europe Cosmetics. *Cosmetics and Personal Care Industry Overview.* Available online: https://www.cosmeticseurope.eu/cosmetics-industry/ (accessed on 28 August 2017).

26. Nunes, M.A.; Rodrigues, F.; Oliveira, M.B.P.P. Grape processing by-products as active ingredients for cosmetic proposes. In *Handbook of Grape Processing By-Products*; Galanakis, C.M., Ed.; Elsevier Inc.: Amsterdam, The Netherlands, 2017; pp. 267–292.

27. Ariede, M.B.; Candido, T.M.; Jacome, A.L.M.; Velasco, M.V.R.; de Carvalho, J.C.M.; Baby, A.R. Cosmetic attributes of algae—A review. *Algal Res.* **2017**, *25*, 483–487. [CrossRef]

28. Thornfeldt, C. *Botanicals. Cosmetic Dermatology*; Wiley-Blackwell: Hoboken, NJ, USA, 2010; pp. 267–280. Available online: https://www.wiley.com/en-us/Cosmetic+Dermatology%3A+Products+and+Procedures-p-9781444359510 (accessed on 25 December 2017).

29. Harnedy, P.A.; FitzGerald, R.J. Bioactive proteins, peptides, and amino acids from macroalgae. *J. Phycol.* **2011**, *47*, 218–232. [CrossRef] [PubMed]

30. Bixler, H.J.; Porse, H. A decade of change in the seaweed hydrocolloids industry. *J. Appl. Phycol.* **2011**, *23*, 321–335. [CrossRef]

31. Food and Agriculture Organization. *The State of World Fisheries and Aquaculture 2014*; Fisheries and Aquaculture Department of the Food and Agricultural Organization of the United Nations: Rome, Italy, 2014.

32. Kolanjinathan, K.; Ganesh, P.; Saranraj, P. Pharmacological importance of seaweeds: A review. *World J. Fish Mar. Sci.* **2014**, *6*, 1–15.

33. Dhargalkar, V.; Pereira, N. Seaweed: Promising plant of the millennium. *Sci. Cult.* **2005**, *71*, 60–66.

34. Pereira, L. A review of the nutrient composition of selected edible seaweeds. In *Seaweed: Ecology, Nutrient Composition and Medicinal Uses*; Pomin, V.H., Ed.; Nova Science Publishers, Inc.: Hauppauge, NY, USA, 2011; pp. 15–47.

35. Baweja, P.; Kumar, S.; Sahoo, D.; Levine, I. Biology of Seaweeds. In *Seaweed in Health and Disease Prevention*; Fleurence, J., Levine, I., Eds.; Elsevier Inc.: Amsterdam, The Netherlands, 2016; pp. 41–106.

36. Wells, M.L.; Potin, P.; Craigie, J.S.; Raven, J.A.; Merchant, S.S.; Helliwell, K.E.; Smith, A.G.; Camire, M.E.; Brawley, S.H. Algae as nutritional and functional food sources: Revisiting our understanding. *J. Appl. Phycol.* **2017**, *29*, 949–982. [CrossRef] [PubMed]

37. Abreu, M.H.; Pereira, R.; Yarish, C.; Buschmann, A.H.; Sousa-Pinto, I. IMTA with *Gracilaria vermiculophylla*: Productivity and nutrient removal performance of the seaweed in a land-based pilot scale system. *Aquaculture* **2011**, *312*, 77–87. [CrossRef]

38. Holdt, S.L.; Kraan, S. Bioactive compounds in seaweed: Functional food applications and legislation. *J. Appl. Phycol.* **2011**, *23*, 543–597. [CrossRef]

39. Ortiz, J.; Uquiche, E.; Robert, P.; Romero, N.; Quitral, V.; Llantén, C. Functional and nutritional value of the Chilean seaweeds *Codium fragile*, *Gracilaria chilensis* and *Macrocystis pyrifera*. *Eur. J. Lipid Sci. Technol.* **2009**, *111*, 320–327. [CrossRef]
40. MacArtain, P.; Gill, C.I.; Brooks, M.; Campbell, R.; Rowland, I.R. Nutritional value of edible seaweeds. *Nutr. Rev.* **2007**, *65*, 535–543. [CrossRef] [PubMed]
41. Pereira, R.; Yarish, C.; Sousa-Pinto, I. The influence of stocking density, light and temperature on the growth, production and nutrient removal capacity of *Porphyra dioica* (Bangiales, Rhodophyta). *Aquaculture* **2006**, *252*, 66–78. [CrossRef]
42. Lee, J.-C.; Hou, M.-F.; Huang, H.-W.; Chang, F.-R.; Yeh, C.-C.; Tang, J.-Y.; Chang, H.-W. Marine algal natural products with anti-oxidative, anti-inflammatory, and anti-cancer properties. *Cancer Cell Int.* **2013**, *13*. [CrossRef] [PubMed]
43. Kılınç, B.; Cirik, S.; Turan, G.; Tekogul, H.; Koru, E. Seaweeds for food and industrial applications. In *Food Industry*; Muzzalupo, I., Ed.; InTech: Rijeka, Croatia, 2013; pp. 735–748.
44. Glicksman, M. Utilization of seaweed hydrocolloids in the food industry. *Hydrobiologia* **1987**, *151*, 31–47. [CrossRef]
45. Personal Care—Global Portfolio. Available online: https://www.cargill.com/doc/1432075967907/personale-care-product-portfolio.pdf (accessed on 24 December 2017).
46. SpecialChem—The Universal Selection Source: Cosmetics Ingredients. Available online: https://cosmetics.specialchem.com/ (accessed on 30 November 2017).
47. SEPPIC. SEPPIC Launches XLYLISHINE™ the New Hair Repairing and Moisturizing Active Ingredient. 2017. Available online: https://worldnews.today/news/25/453/seppic-launches-a-new-hair-repairing-and-moisturizing-active-ingredient.html (accessed on 22 December 2017).
48. Bowe, W.P.; Pugliese, S. Cosmetic benefits of natural ingredients. *J. Drugs Dermatol.* **2014**, *13*, 1021–1025. [PubMed]
49. Agatonovic-Kustrin, S.; Morton, D. Cosmeceuticals derived from bioactive substances found in marine algae. *Oceanography* **2013**, *1*, 1–11. [CrossRef]
50. Ngo, D.-H.; Wijesekara, I.; Vo, T.-S.; Van Ta, Q.; Kim, S.-K. Marine food-derived functional ingredients as potential antioxidants in the food industry: An overview. *Food Res. Int.* **2011**, *44*, 523–529. [CrossRef]
51. Athukorala, Y.; Kim, K.-N.; Jeon, Y.-J. Antiproliferative and antioxidant properties of an enzymatic hydrolysate from brown alga, *Ecklonia cava*. *Food Chem. Toxicol.* **2006**, *44*, 1065–1074. [CrossRef] [PubMed]
52. Nepalia, A.; Singh, A.; Mathur, N.; Pareek, S. An overview of the harmful additives and contaminants possibly present in baby cosmetic products. *Int. J. Chem. Sci.* **2017**, *15*, 127.
53. Jo, J.-H.; Kim, D.; Lee, S.; Lee, T.K. Total phenolic contents and biological activities of Korean seaweed extracts. *Food Sci. Biotechnol.* **2005**, *14*, 798–802.
54. Da Costa, E.; Melo, T.; Moreira, A.; Bernardo, C.; Helguero, L.; Ferreira, I.; Cruz, M.; Rego, A.; Domingues, P.; Calado, R.; et al. Valorization of lipids from *Gracilaria* sp. through lipidomics and decoding of antiproliferative and anti-inflammatory activity. *Mar. Drugs* **2017**, *15*. [CrossRef] [PubMed]
55. Pérez, M.J.; Falqué, E.; Domínguez, H. Antimicrobial action of compounds from marine seaweed. *Mar. Drugs* **2016**, *14*, 52. [CrossRef] [PubMed]
56. De Almeida, C.L.F.; Falcão, D.S.; Lima, D.M.; Gedson, R.; Montenegro, D.A.; Lira, N.S.; De Athayde-Filho, P.F.; Rodrigues, L.C.; de Souza, M.D.V.; Barbosa-Filho, J.M.; et al. Bioactivities from marine algae of the genus Gracilaria. *Int. J. Mol. Sci.* **2011**, *12*, 4550–4573. [CrossRef] [PubMed]
57. Le Lann, K.; Surget, G.; Couteau, C.; Coiffard, L.; Cérantola, S.; Gaillard, F.; Larnicol, M.; Zubia, M.; Guérard, F.; Poupart, N.; et al. Sunscreen, antioxidant, and bactericide capacities of phlorotannins from the brown macroalga *Halidrys siliquosa*. *J. Appl. Phycol.* **2016**, *28*, 3547–3559. [CrossRef]
58. Siahaan, E.A.; Pangestuti, R.; Munandar, H.; Kim, S.-K. Cosmeceuticals properties of Sea Cucumbers: Prospects and trends. *Cosmetics* **2017**, *4*, 26. [CrossRef]
59. Hardouin, K.; Burlot, A.-S.; Umami, A.; Tanniou, A.; Stiger-Pouvreau, V.; Widowati, I.; Bedoux, G.; Bourgougnon, N. Biochemical and antiviral activities of enzymatic hydrolysates from different invasive French seaweeds. *J. Appl. Phycol.* **2014**, *26*, 1029–1042. [CrossRef]
60. De Souza, M.C.R.; Marques, C.T.; Dore, C.M.G.; da Silva, F.R.F.; Rocha, H.A.O.; Leite, E.L. Antioxidant activities of sulfated polysaccharides from brown and red seaweeds. *J. Appl. Phycol.* **2007**, *19*, 153–160. [CrossRef] [PubMed]

61. Fleita, D.; El-Sayed, M.; Rifaat, D. Evaluation of the antioxidant activity of enzymatically-hydrolyzed sulfated polysaccharides extracted from red algae *Pterocladia capillacea*. *Food Sci. Technol.* **2015**, *63*, 1236–1244. [CrossRef]

62. Zhang, Z.; Wang, F.; Wang, X.; Liu, X.; Hou, Y.; Zhang, Q. Extraction of the polysaccharides from five algae and their potential antioxidant activity in vitro. *Carbohydr. Polym.* **2010**, *82*, 118–121. [CrossRef]

63. Leelapornpisid, P.; Mungmai, L.; Sirithunyalug, B.; Jiranusornkul, S.; Peerapornpisal, Y. A novel moisturizer extracted from freshwater macroalga [*Rhizoclonium hieroglyphicum* (C.Agardh) Kützing] for skin care cosmetic. *Chiang Mai J. Sci.* **2014**, *41*, 1195–1207.

64. Choi, J.-S.; Moon, W.S.; Choi, J.N.; Do, K.H.; Moon, S.H.; Cho, K.K.; Han, C.-J.; Choi, I.S. Effects of seaweed *Laminaria japonica* extracts on skin moisturizing activity in vivo. *J. Cosmet. Sci.* **2013**, *64*, 193–209. [PubMed]

65. Bedoux, G.; Hardouin, K.; Burlot, A.S.; Bourgougnon, N. Bioactive components from seaweeds: Cosmetic applications and future development. *Adv. Bot. Res.* **2014**, *71*, 345–378.

66. Chojnacka, K.; Saeid, A.; Witkowska, Z.; Tuhy, L. Biologically active compounds in seaweed extracts—The prospects for the application. *Open Conf. Proc. J.* **2012**, *3*, 20–28. [CrossRef]

67. Fleurence, J.; Ar Gall, E. Antiallergic Properties. In *Seaweed in Health and Disease Prevention*; Elsevier Inc.: Amsterdam, The Netherlands, 2016; pp. 389–406.

68. Dumay, J.; Morançais, M. Proteins and Pigments. In *Seaweed in Health and Disease Prevention*; Fleurence, J., Levine, I., Eds.; Elsevier Inc.: Amsterdam, The Netherlands, 2016; pp. 275–318.

69. Sá, A.G.A.; Meneses, A.C.; Araújo, P.H.H.; Oliveira, D. A review on enzymatic synthesis of aromatic esters used as flavor ingredients for food, cosmetics and pharmaceuticals industries. *Trends Food. Sci. Technol.* **2017**, *69*, 95–105. [CrossRef]

70. Fleurence, J. Seaweed proteins. *Trends Food Sci. Technol.* **1999**, *10*, 25–28. [CrossRef]

71. Marinho-Soriano, E.; Fonseca, P.; Carneiro, M.; Moreira, W. Seasonal variation in the chemical composition of two tropical seaweeds. *Bioresour. Technol.* **2006**, *97*, 2402–2406. [CrossRef] [PubMed]

72. Samarakoon, K.; Jeon, Y.-J. Bio-functionalities of proteins derived from marine algae—A review. *Food Res. Int.* **2012**, *48*, 948–960. [CrossRef]

73. Malerich, S.; Berson, D. Next generation Cosmeceuticals. *Dermatol. Clin.* **2014**, *32*, 13–21. [CrossRef] [PubMed]

74. Kang, H.K.; Seo, C.H.; Park, Y. Marine peptides and their anti-infective activities. *Mar. Drugs* **2015**, *13*, 618–654. [CrossRef] [PubMed]

75. Schagen, S.K. Topical peptide treatments with effective anti-aging results. *Cosmetics* **2017**, *4*, 16. [CrossRef]

76. Cardozo, K.H.; Marques, L.G.; Carvalho, V.M.; Carignan, M.O.; Pinto, E.; Marinho-Soriano, E.; Colepicolo, P. Analyses of photoprotective compounds in red algae from the Brazilian coast. *Braz. J. Pharmacogn.* **2011**, *21*, 202–208. [CrossRef]

77. Pandey, A.; Pandey, S.; Pathak, J.; Ahmed, H.; Singh, V.; Singh, S.P.; Sinha, R.P. Mycosporine-Like Amino Acids (MAAs) Profile of Two Marine Red Macroalgae, *Gelidium* sp. and *Ceramium* Sp. *Int. J. Appl. Biotechnol. Biochem.* **2017**, *5*, 12–21. [CrossRef]

78. Kumari, P.; Kumar, M.; Gupta, V.; Reddy, C.; Jha, B. Tropical marine macroalgae as potential sources of nutritionally important PUFAs. *Food Chem.* **2010**, *120*, 749–757. [CrossRef]

79. Misurcova, L.; Ambrozova, J.; Samek, D. Seaweed lipids as nutraceuticals. *Adv. Food Nutr. Res.* **2011**, *64*, 339–355. [PubMed]

80. Tapiero, H.; Nguyen Ba, G.; Couvreur, P.; Tew, K.D. Polyunsaturated fatty acids (PUFA) and eicosanoids in human health and pathologies. *Biomed. Pharmacother.* **2002**, *56*, 215–222. [CrossRef]

81. Osman, N.A.R.; Abdo, B.; Mohamed, S.E.-T. Assessment of the nutritional value and native agar content of the red alga *Gracilaria foliifera* (Forsskal) Borgesen from the Red Sea coast of Sudan. *J. Algal Biomass Utln.* **2017**, *8*, 48–63.

82. Kendel, M.; Wielgosz-Collin, G.; Bertrand, S.; Roussakis, C.; Bourgougnon, N.; Bedoux, G. Lipid composition, fatty acids and sterols in the seaweeds *Ulva armoricana*, and *Solieria chordalis* from Brittany (France): An analysis from nutritional, chemotaxonomic, and antiproliferative activity perspectives. *Mar. Drugs* **2015**, *13*, 5606–5628. [CrossRef] [PubMed]

83. Hegazi, M.M. Separation, identification and quantification of photosynthetic pigments from three red sea seaweeds using reversed-phase high-performance liquid chromatography. *Egypt. J. Biol.* **2002**, *4*, 1–6.

84. Belitz, H.-D.; Grosch, W.; Schieberle, P. *Lipids. Food Chemistry*; Springer: Berlin, Germany, 2009; pp. 158–247.

85. Lyons, N.M.; O'Brien, N.M. Modulatory effects of an algal extract containing astaxanthin on UVA-irradiated cells in culture. *J. Dermatol. Sci.* **2002**, *30*, 73–84. [CrossRef]

86. Kelman, D.; Posner, E.K.; McDermid, K.J.; Tabandera, N.K.; Wright, P.R.; Wright, A.D. Antioxidant activity of Hawaiian marine algae. *Mar. Drugs* **2012**, *10*, 403–416. [CrossRef] [PubMed]

87. Aneiros, A.; Garateix, A. Bioactive peptides from marine sources: Pharmacological properties and isolation procedures. *J. Chromatogr. B* **2004**, *803*, 41–53. [CrossRef] [PubMed]

88. Sonani, R.R.; Singh, N.K.; Kumar, J.; Thakar, D.; Madamwar, D. Concurrent purification and antioxidant activity of phycobiliproteins from *Lyngbya* sp. A09DM: An antioxidant and anti-aging potential of phycoerythrin in *Caenorhabditis elegans*. *Process Biochem.* **2014**, *49*, 1757–1766. [CrossRef]

89. Guillerme, J.-B.; Couteau, C.; Coiffard, L. Applications for marine resources in cosmetics. *Cosmetics* **2017**, *4*, 35. [CrossRef]

90. Lecas, S.; Boursier, E.; Fitoussi, R.; Vié, K.; Momas, I.; Seta, N.; Achard, S. In vitro model adapted to the study of skin ageing induced by air pollution. *Toxicol. Lett.* **2016**, *259*, 60–68. [CrossRef] [PubMed]

91. Farris, P.K. Natural ingredients and their applications in dermatology. In *Practical Dermatology*; Rick Ehrlich: New York, NY, USA, 2010; pp. 51–54.

92. Pillai, S.; Cornell, M.; Oresajo, C. Part 1: Skin Physiology Pertinent to Cosmetic Dermatology. Epidermal barrier. In *Cosmetic Dermatology—Products and Procedures*; Draelos, Z.D., Ed.; Blackwell Publishing: Hoboken, NJ, USA, 2010; pp. 3–12.

93. Wang, J.; Jin, W.; Hou, Y.; Niu, X.; Zhang, H.; Zhang, Q. Chemical composition and moisture-absorption/retention ability of polysaccharides extracted from five algae. *Int. J. Biol. Macromol.* **2013**, *57*, 26–29. [CrossRef] [PubMed]

94. Robert, L.; Labat-Robert, J.; Robert, A.-M. Physiology of skin aging. *Pathol. Biol.* **2009**, *57*, 336–341. [CrossRef] [PubMed]

95. Shibata, T.; Fujimoto, K.; Nagayama, K.; Yamaguchi, K.; Nakamura, T. Inhibitory activity of brown algal phlorotannins against hyaluronidase. *Int. J. Food Sci. Technol.* **2002**, *37*, 703–709. [CrossRef]

96. Heo, S.-J.; Jeon, Y.-J. Protective effect of fucoxanthin isolated from *Sargassum siliquastrum* on UV-B induced cell damage. *J. Photochem. Photobiol. B* **2009**, *95*, 101–107. [CrossRef] [PubMed]

97. Kindred, C.; Halder, R.M. Pigmentation and skin of color. In *Cosmetic Dermatology*; Wiley-Blackwell: Hoboken, NJ, USA, 2010; pp. 27–37.

98. Cha, S.H.; Ko, S.C.; Kim, D.; Jeon, Y.J. Screening of marine algae for potential tyrosinase inhibitor: Those inhibitors reduced tyrosinase activity and melanin synthesis in zebrafish. *J. Dermatol.* **2011**, *38*, 354–363. [CrossRef] [PubMed]

99. Heo, S.-J.; Ko, S.-C.; Kang, S.-M.; Cha, S.-H.; Lee, S.-H.; Kang, D.-H.; Jung, W.-K.; Affan, A.; Oh, C.; Jeon, Y.-J. Inhibitory effect of diphlorethohydroxycarmalol on melanogenesis and its protective effect against UV-B radiation-induced cell damage. *Food Chem. Toxicol.* **2010**, *48*, 1355–1361. [CrossRef] [PubMed]

100. Senevirathne, M.; Ahn, C.-B.; Je, J.-Y. Enzymatic extracts from edible red algae, *Porphyra tenera*, and their antioxidant, anti-acetylcholinesterase, and anti-inflammatory activities. *Food Sci. Biotechnol.* **2010**, *19*, 1551–1557. [CrossRef]

101. Rahimpour, Y.; Hamishehkar, H. Liposomes in cosmeceutics. *Expert Opin. Drug Deliv.* **2012**, *9*, 443–455. [CrossRef] [PubMed]

102. Al-Bader, T.; Byrne, A.; Gillbro, J.; Mitarotonda, A.; Metois, A.; Vial, F.; Rawlings, A.V.; Laloeuf, A. Effect of cosmetic ingredients as anticellulite agents: Synergistic action of actives with in vitro and in vivo efficacy. *J. Cosmet. Dermatol.* **2012**, *11*, 17–26. [CrossRef] [PubMed]

103. Jang, W.S.; Choung, S.Y. Antiobesity effects of the ethanol extract of *Laminaria japonica* Areshoung in high-fat-diet-induced obese rat. *J. Evid. Based Complement. Altern. Med.* **2013**, *2013*, 492807.

104. Westby, T.; Cadogan, A.; Duignan, G. In vivo uptake of iodine from a *Fucus serratus* Linnaeus seaweed bath: Does volatile iodine contribute? *Environ. Geochem. Health* **2017**, 1–9. [CrossRef] [PubMed]

105. Berardesca, E.; Abril, E.; Rona, C.; Vesnaver, R.; Cenni, A.; Oliva, M. An effective night slimming topical treatment. *Int. J. Cosmet. Sci.* **2012**, *34*, 263–272. [CrossRef] [PubMed]

![cosmetics logo] *cosmetics*

MDPI

Review

Coffee Silverskin: A Review on Potential Cosmetic Applications

Sílvia M. F. Bessada *, Rita C. Alves * and M. Beatriz P. P. Oliveira

REQUIMTE/LAQV, Department of Chemical Sciences, Faculty of Pharmacy, University of Porto, 4050-313 Porto, Portugal; beatoliv@ff.up.pt
* Correspondence: silviabessada@gmail.com (S.M.F.B.); rita.c.alves@gmail.com (R.C.A.);
  Tel.: +351-220-428-640 (S.M.F.B. & R.C.A.)

Received: 7 November 2017; Accepted: 28 December 2017; Published: 3 January 2018

Abstract: Coffee silverskin, the major coffee-roasting by-product, is currently used as fuel and for soil fertilization. However, there are several studies reporting silverskin as a good source of bioactive compounds that can be extracted and further used by cosmetic industry. Its high antioxidant potential may be due to the synergistic interaction of chlorogenic acids (1–6%), caffeine (0.8–1.25%), and melanoidins (17–23%), among other antioxidant compounds. The bioactive compounds of silverskin can answer to the new fields of cosmetic industry on natural active ingredient resources that improve health skin appearance, counteract skin aging and related diseases, in an environmentally friendly approach. Skin aging is a complex process associated with oxidative metabolism and reactive oxygen species (ROS) generation. ROS production increase matrix metalloproteinases (MMPs), as well as pro-inflammatory mediators, resulting in consequent skin damage and aging. To counteract this process, cosmetic industry is looking for compounds able to increase MMP inhibitory activities, hyaluronidase inhibitory activity, expression of collagen and elastase inhibitory activity, as potential bioactive ingredients with anti-aging purposes. This review focuses on skin aging factors and the potential anti-aging, anti-inflammatory, antimicrobial, anti-cellulite and anti-hair loss activity, as well as protection against UV damage, of coffee silverskin and their bioactive compounds.

Keywords: coffee silverskin; by-product; anti-aging; antioxidant; anti-inflammatory; antimicrobial; anti-cellulite; anti-hair loss; skin damage protection

---

1. Introduction

Coffee is the most popular beverage over the world, being its regular consumption highly increasing. The main coffee production is based on two plant species: *Coffea arabica* and *Coffea canephora*, also known as arabica and robusta coffees, respectively. Along the several steps of coffee production, a huge amount of residue is generated (e.g., husks, hulls, defective beans, coffee silverskin, and spent coffee grounds). Indeed, coffee wastes and by-products constitute a source of severe contamination and an environmental problem. The chemical characterization of coffee residues and their use as potential bioactive ingredients for development of novel functional products emerge as a factor of environmental sustainability and economic recovery to the companies of coffee processing and roasting. Coffee silverskin (CS) is a thin tegument that directly covers the coffee seed. During the roasting process, coffee beans expand and this thin layer is detached, becoming the main by-product of coffee roasting industries [1]. CS, compared to other coffee by-products, is a relatively stable product due to its lower moisture content (5–7%) [2,3]. Currently used as fuel, for composting and soil fertilization, CS represents a good source of several bioactive compounds that can be extracted and further used for food, cosmetic and pharmaceutical purposes [4].

CS is composed by a high amount of dietary fiber (56–62%), especially soluble fiber (~87%) [2,3]. It contains cellulose (18%) and hemicellulose (13%), being this last composed by xylose (4.7%),

---

arabinose (2.0%), galactose (3.8%), and mannose (2.6%) [5]. It is also rich in protein (19%) and minerals (8% ash). Mineral composition consists mainly in potassium, magnesium and calcium (~5 g, 2 g, and 0.5 g per 100 g silverskin, respectively) [4]. Fat content varies from 1.6% to 3.3%, being dependent on the geographical origin of the coffee. Triacylglycerols are the major components (48%), followed by free fatty acids (21%), esterified sterols (15%), free sterols (13%), and diacylglycerols (4%) [6,7]. According to a previous study [4], CS presents mainly saturated fatty acids (65%), followed by polyunsaturated (28%) and monounsaturated (7%) ones. C18:2n6c was the major fatty acid found (24%), followed by C16:0 (22%), C22:0 (15%), and C20:0 (14%). Regarding antioxidant compounds, Costa et al. [4] analyzed, for the first time, the vitamin E profile of CS, which presents a total content around 4.17 mg/100 g. Four tocopherols (α, β, γ, and δ) and three tocotrienols (β, γ, and δ) were the vitamers found, being α-tocopherol (2.25 mg/100 g) the major one, followed by β-tocotrienol (0.95 mg/100 g) [4]. Other important bioactive compounds present in CS are chlorogenic acids (1–6%) (being 5-*O*-, 3-*O*- and 4-*O*-caffeoylquinic acids the most relevant ones), caffeine (0.8–1.25%), and Maillard reaction products formed during the roasting process, namely melanoidins (17–23%). Indeed, chlorogenic acids and their thermal degradation products seem to be involved in the formation of melanoidins, along with others compounds, such as polysaccharides (galactomannans and arabinogalactans) and proteins [4,8,9].

Attending nutritional and chemical composition, antioxidant activity and several bioactive compounds, CS emerges as a particularly interesting product to be used in food, cosmetic and pharmaceutical industries. However, the use of this coffee waste must take into account safety concerns, since CS could contain ochratoxin A (OTA), an *Aspergillus ochraceus* and *Penicillium verrucosum* toxin, classified by the International Agency for Research on Cancer (IARC) as a possible human carcinogen. The Commission Regulation (EC) No 123/2005 defined OTA limits as 5 µg/kg for roasted coffee and 10 µg/kg for soluble coffee. For CS, there is no specific OTA regulation limit [10,11]. Some authors have described the effect of coffee roasting (time/temperature) on the reduction of OTA levels (reduction > 90%), reporting a level below 4 µg/kg in CS [2,12]. Nevertheless, Toschi et al. [7] quantified OTA levels between 18.7 and 34.4 ug/kg CS, corresponding to values three times higher than those fixed by the European Commission. The same study also describes the presence of phytosterol oxidation products (114.11 mg/100 g); therefore, it is crucial to set up good practices of production and a rigorous quality control, developing suitable methods to minimize the presence of these undesirable compounds.

Until now, several innovative approaches have been suggested for CS, essentially based on its richness in dietary fiber, phenolic compounds and other antioxidants, such as melanoidins [2,3,13,14]. For instance, Mussato et al. [14] suggested its incorporation in flakes, breads, biscuits and snacks. In addition, Pourfarzad et al. [15] used this by-product to improve quality, shelf life, and sensorial properties of Barbari flat bread, while reducing its caloric density and increasing the dietary fiber content. In turn, Martinez-Saez et al. [16] used CS to prepare a novel antioxidant beverage for body weight control, while new approaches on cosmetic uses have been reported by Rodrigues and colleagues [17,18].

Cosmetic industry has been looking for new active ingredients, due to the consumer demand for more natural and environmentally friendly products obtained by sustainable resources that improve health skin appearance. Silverskin is a potential candidate to replace synthetic chemicals as active ingredients in cosmetic formulations due to their high antioxidant potential, phenolic compounds, melanoidins and caffeine contents [19]. Table 1 shows the major CS bioactive compounds that can be extracted for cosmetic purposes, based on their biological activities.

Table 1. Major CS bioactive compounds and biological activities (Adapted from Rodrigues et al. [19]).

Compound	Biological Activities	References
Caffeine	Antioxidant and anti-aging activity; Thermogenic and anti-cellulite activity; Protection against UV damage; Increase of blood circulation in the skin; Inhibition of 5α-reductase and hyaluronidase activities.	[7,19,20]
Caffeoylquinic acids/ Feruloylquinic acids/ p-coumaroylquinic acids	Anti-aging activity; Protection against UV damage; Antimicrobial and anti-inflammatory activity.	[17,21]
Melanoidins	Antioxidant, anti-aging, antimicrobial and anti-inflammatory activities.	[22]

2. Skin Aging and Related Diseases

Skin, the largest organ of human body, has multiples functions, being the protection against environmental factors one of the most important. As a living organ, skin goes through significant changes throughout a person's lifetime, so the appearance of a luminous visual and a healthy skin, less aged or damaged is a constant consumer requirement and a challenge for cosmetic products. In fact, the search for new and improved multifunctional ingredients from natural sources has been a relevant issue in cosmetic fields.

Aging is defined as an accumulation of changes in the cells, tissues or organs over the time, which leads to a progressive loss of structure and function, an inevitable, universal phenomenon that depends on each person's genetic capital and lifestyle. Skin aging is a slow and complex process including intrinsic and extrinsic mechanisms involved, which induce skin changes as thinning, dryness, laxity, fragility, enlarged pores, fine lines, and wrinkles [23,24]. Intrinsic aging occurs as a natural consequence of human physiological and genetic changes, while extrinsic aging is mainly caused by cumulative exposure to external harmful factors such as UV radiation, atmospheric pollution, and infectious agents that induce alterations and skin damage. Oxidative stress, the major cause of skin accelerated aging and related diseases, can be defined as the imbalance between ROS and reactive nitrogen species (RNS) (e.g., superoxide anion radical, hydroxyl, alkoxyl and lipid peroxyl radicals, nitric oxide and peroxynitrite) and antioxidants. The human body developed a complex endogenous antioxidant system to counteracted "physiologic" oxidative stress, which includes endogenous antioxidant enzymes such as superoxide dismutase, catalase, glutathione peroxidase, and non-enzymatic compounds (e.g., glutathione, proteins, coenzyme Q, and lipoic acid), as well as exogenous antioxidants obtained from diet (vitamin C and E, carotenoids, phenolic compounds, etc.) [25]. In fact, different species and plant constituents have been chemically characterized as natural sources of antioxidant compounds with a high potential of application [26]. When cells are subjected to excessive levels of oxidants or the depletion of antioxidants, oxidative stress occurs. Under normal conditions, oxidants are natural by-products of physiological processes (e.g., mitochondria, peroxisomes and plasma membrane), which have positive physiological effects on cells, being involved in differentiation, proliferation, and signal transduction. However, reactive species can also be generated by exogenous sources and cause DNA, protein and lipid damages. With age, an imbalance arises between oxidants production and antioxidants (created or ingested), which can lead to skin aging and some related diseases (e.g., dermatitis, sunburn, eczema, vasculitis, cancer, etc.). Indeed, skin aging seems to be associated to oxidative metabolism and reactive species generation [19,23,27].

Keratinized stratified epidermis and an underlying thick layer of collagen-rich dermal connective tissue are important components of the young skin. Summarizing, skin is composed by three important layers: the epidermis keratinization process leads to a thin *stratum corneum* layer that provide skin protection, which, together with a hydrolipidic film, create a true barrier against external stress; the internal thick layer of skin, dermis, is comprised by water, elastin and collagen fibers, being the supportive tissue; and the hypodermis is the deepest and thickest layer of skin, rich in adipose cells and protects the body from physical shock and temperature variations. It is also an energy reserve [28].

The loss of structural protein (type-1 collagen) in the dermal layer, is the major cause of wrinkle-aging process formation. Indeed, skin aging is related to the reduction of collagen production and increased levels of matrix metalloproteinases (MMP), mostly MMP-1 and elastase enzymes, which are responsible for the collagen and elastin breakdown. In addition, molecular mechanisms of skin aging can induce inflammatory responses, which, in turn, results in upregulation of metalloproteinases (MMP-1, MMP-3, and MMP-9), contributing to the degradation of skin collagen and connective tissue [23,24,28].

2.1. Anti-Aging Activity

As human beings age, the skin thins, dries, wrinkles and becomes pigmented. Skin aging is related to collagen fibers fragmentation by MMP-1 and increased mitochondrial ROS production and oxidative stress, resulting in common deletions of mitochondrial DNA. ROS production is known to cause the activation of: (i) AP-1, the transcription factor responsible for MMPs production; and (ii) nuclear transcription factor-kappa B (NF-kB), which is essential in normal physiology, however, an inappropriate regulation has been associated with several chronic diseases, inflammation and cancer [23,28]. NF-kB upregulates the transcription of pro-inflammatory mediators, such interleukin-1 (IL-1), IL-6, and IL-8, and tumor necrosis factor-α (TNF-α). These pro-inflammatory mediators, in a positive feedback loop, stimulate further production of ROS and activate AP-1 and NF-kB, resulting in more damage and skin aging [19,27,29].

As previously reported, CS have a high antioxidant content, especially chlorogenic acids and caffeine, two promising compounds for anti-wrinkle products. Recent research was developed on hyaluronidase inhibitory activity of CS extracts [20]. During aging, hyaluronic acid content diminishes and skin becomes dry and wrinkled. Indeed, hyaluronidase degrades hyaluronic acid, lowering its viscosity, increasing permeability, and leading to extracellular matrix (collagen and elastin fibers) destruction [30]. According to Furusawa et al. [20], the higher molecular-weight substances present in CS extracts could contribute to the hyaluronidase inhibition effect, with acidic polysaccharides (mainly composed by uronic acid) appearing to play a major role. In the cosmetics point of view, Rodrigues and colleagues [17] reported the skin compatibility and safety of CS extracts. In another study, Rodrigues et al. [18] performed an in vivo evaluation of hyaluronidase inhibitory effect. This assay was performed on 20 human volunteers, using a coffee silverskin-based cream, twice a day, over 28 days. After this period, the hydration and firmness was evaluated by comparing the cream containing CS with a formulation supplemented with 1.5% of HyaCare® Filler CL (a cross-linked polysaccharide made from fermentation-derived hyaluronic acid). CS was shown to be an effective ingredient, with similar results to hyaluronic acid, in the improvement of skin hydration and firmness [18]. In two other studies [31,32], a body and a hand cream containing CS extract were developed. For both, antioxidant activity and consumer acceptability was high, and a viability cell decrease (in keratinocytes and fibroblasts) was not observed.

A new approach on anti-aging effect of CS extracts against accelerated aging caused by oxidative agents, was performed by Iriondo-DeHond et al. [33]. In this study, accelerated aging in *C. elegans* (animal model) was induced by ultraviolet radiation C (UVC) and in HaCaT cells (skin model) using *tert*-butyl hydroperoxide (*t*-BOOH). The nematodes *C. elegans* treated with CS extract (1 mg/mL) showed a significant increased longevity compared to those cultured on a standard diet. The increased longevity observed was similar to that of the nematodes fed with chlorogenic acid or vitamin C (0.1 μg/mL). On the other hand, the tested concentrations of CS extracts were not cytotoxic, and a CS extract of 1 mg/mL gave resistance to skin cells when oxidative damage was induced by t-BOOH. According to the authors, the anti-aging properties of CS may be due to its complex mixture of antioxidants that act in a synergistic combination [33]. In sum, CS extracts have the potential to be used as an ingredient in skin cosmetic product to reduce the production of intracellular ROS in keratinocytes and improving skin health. Additionally, as will be discussed later in Section 2.4, CS extracts protect against skin photoaging induced by UV radiation.

2.2. Anti-Inflammatory Activity

Skin inflammation can be defined as a skin response to an injury, infection or destruction, normally characterized by heat, redness, pain, swelling or disturbed skin physiological functions. ROS contribute to pro-inflammatory signaling cascades and consequent production of cytokines such as interleukin-1β (IL-1β), and TNF-α [29]. Importantly, there is reciprocal activation by oxidative stress and inflammatory mediators, since cytokines and TNF-α also induce $O_2^{\bullet-}$ production, activate or induce the expression of inflammatory enzymes (NOX2 and cyclooxygenase) and trigger NADPH oxidase-dependent inflammation. In addition, increase collagenase and/or elastase enzymes expression, decreasing the tensile strength and elasticity of the skin. Therefore, ROS, cytokines and interleukins act in synergism during the inflammatory skin process, inducing keratinocyte proliferation and monocyte recruitment to the injury side [34,35].

Recent studies have evaluated the anti-inflammatory effect of chlorogenic acid (and its metabolite caffeic acid), a phenolic compound prevalent in CS composition [36]. In this study, the IL-8 production in a human intestinal cell line (Caco-2) was induced by combined stimulation with TNF-α and H_2O_2. The results demonstrated that cholorogenic acid and caffeic acid inhibit the induced IL-8 production, suggesting an important anti-inflammatory effect [36]. In turn, Hwang et al. [37] reported the caffeine anti-inflammatory effect on lipopolysaccharide (LPS)-induced inflammation. In that study, RAW264.7 cells were treated with several concentrations of caffeine in the presence or absence of LPS. The results showed a decrease in LPS-induced inflammatory mediators by regulating NF-κB activation [37].

Melanoidins are bioactive compounds that are formed during the coffee roast [22], and have also been found in CS (17–23%) [9]. Important bioactive properties have been reported to this group of compounds as antioxidant activity, inhibition of MMPs, antimicrobial and anti-inflammatory activities [22]. Until now, only studies on coffee beans and coffee-based beverages have reported the melanoidins anti-inflammatory effect. Paur et al. [29] showed that extracts of dark-roasted coffee inhibit NF-κB activity by more than 80%, in LPS-induced NF-κB activation. Similar evidences were reported by Vitaglione et al. [38] on reducing TNF-α expression.

2.3. Antimicrobial Activity

Rodrigues et al. [21] demonstrated the antimicrobial activity of CS extracts against pathogenic bacteria such as *Staphylococcus aureus* (ATCC 6538 and MRSA), *Staphylococcus epidermidis*, *Escherichia coli* (ATCC 1576 and MRSA) and *Klebsiella pneumoniae* (ATCC4352). In another study, Jiménez-Zamora and colleagues [39] show the prebiotic, antimicrobial and antioxidant capacity of CS, and highlighted the antimicrobial activity of melanoidins extracted from spent coffee grounds. Depending on their concentration, coffee melanoidins can have bacteriostatic or bactericide action. Their antimicrobial activity has been widely studied by different research groups [21,40]. In addition, other coffee compounds, naturally present in CS, as chlorogenic acids and caffeine [4], could synergistically interact for antibacterial activity against *S. mutans*, a microorganism related with dental plaque formation [41]. In sum, CS could have a promising application in skin infection diseases, or even as a preservative for final cosmetic formulations [19,40].

2.4. Protection against Skin UV Damage

As previously reported in Section 2.2, skin inflammation can be a consequence of deleterious effects of extrinsic factors, being overexposure to UV solar radiation (mainly UVB) an important factor to skin-related disorders. UVB induces overproduction of ROS such as superoxide anion ($O_2^{\bullet-}$) and singlet O_2 ($^1O_2$), which are critical events for the onset of oxidative stress conditions. For instance, $O_2^{\bullet-}$ reacts with hydrogen peroxide (H_2O_2) generating the cytotoxic hydroxyl radical ($^\bullet OH$). In turn, $^\bullet OH$ causes lipid peroxidation, a well-established detrimental consequence of UVB chronic exposure, evidenced by erythema, edema, hyperpigmentation, premature skin aging, and cancer. Exposure to UV light has an adverse effect on skin components such as elastin and collagen [23], eventually leading to formation of wrinkles. The formation of cholesterol oxidation products through UV exposure (by generation of an excited singlet oxygen that reacts with the double bond in position 5,6 of the B ring, following similar oxidative pathways as monounsaturated fatty acids) has also been reported [42].

In fact, increased ROS production induces inflammatory cytokines secretion and enhances dermal fibroblast MMPs levels, and decrease the procollagen synthesis. Expression of MMPs leads to degradation of extracellular matrix proteins (e.g., collagen and elastin fibers, fibronectin, and laminin), resulting in skin elasticity decrease and wrinkle formation [34]. Considering the synergic effect of ROS and inflammatory mediators, the use of compounds with antioxidant and anti-inflammatory effects are a promising approach to inhibit UVB irradiation-induced skin damage. In this context, a high attention has been paid to antioxidants from natural sources and their advantages (i.e., naturally occurring agents are considered to be less toxic and can be more economic). In this context, a new approach on sunscreens based in natural sources of antioxidants and anti-inflammatory compounds have been under development. Kitagawa et al. [43] evaluated the chlorogenic acid activity on skin protection against UV-induced damage. An oil/water product containing this compound was incubated on excised guinea pig dorsal skin and Yucatan micro pigskin. Antioxidant activity of chlorogenic acid was observed by ROS reduction, suggesting the potential of this compound to protect skin against UV-induced oxidative damage. While there are more than 25 types of MMPs, UV light exposure is particularly associated with induction of MMP-1, MMP-3, and MMP-9, which play critical roles in the skin aging process [44]. Recently, the potential effect of compounds isolated from *Coffea arabica* beans against UV-B induced skin damage was evaluated by Cho et al. [45]. The results evidenced individual anti-wrinkle effects of chlorogenic acid, pyrocatechol, and 3,4,5-tricaffeoyl quinic acid against UV-B stimulated mouse fibroblast cells (CCRF), by measuring expression levels of MMP-1, MMP-3, MMP-9, and type-I procollagen. Chlorogenic acid effectively suppressed the expression of the MMP-1, MMP-3, and MMP-9 and has potential to upregulate procollagen synthesis in UV-induced fibroblasts, which could be effective to reduce photoaging that leads to wrinkle formation. Additionally, chlorogenic acid presents a good sun protection factor (SPF)—a factor that measures the capacity of a compound to absorb UV radiation—and in vitro DNA protective effect. This study indicates that this phenolic compound has potential to be used as a preventive agent against premature skin aging induced by UV radiation [45]. Although this study was performed in coffee beans, chlorogenic acid is also the major phenolic in CS (1–6%).

Caffeine is another natural compound of CS with UV-absorption properties. Lu and colleagues [46] studied the effect of caffeine topical application to prevent UVB-induced skin-cancer development on SKH-1 hairless mice. The mouse models were treated topically with caffeine once a day (five days/week) for 18 weeks, and the results showed that topical applications of caffeine decreased the number of nonmalignant and malignant skin tumors per mouse (44% and 72%, respectively). In another study, Koo and collaborators [47] evaluated the effect of topical application of caffeine in SKH-1 hairless mice skin after being exposed to UVB (three times/week for 11 weeks). Caffeine application promoted the deletion of DNA-damaged keratinocytes, suggesting that it may diminish photodamage and photocarcinogenesis. In addition, topical application of caffeine 30 minutes before a high dose UVB exposure showed a protective effect against sunburn lesions [47]. Recently, Choi et al. [48] evaluated the protective action of oil fraction (containing 547.32 mg/mL of caffeine and 119.25 mg/mL of chlorogenic acid) and ethanolic extracts (42.58 and 50.75 mg/mL, respectively) of spent coffee grounds (from instant coffee). Topical application of the oil or ethanol fraction significantly reduced the UVB-induced wrinkle formation in mice dorsal skin. In addition, the combined application of oil and ethanolic fraction led to a decrease in the wrinkle area by over 35%, epidermal thickness (40%), transdermal water loss (27%) and erythema formation (48%) that result from UVB exposure, when compared with the UVB-treated control. In addition, polarization-sensitive optical coherence tomography and antibody-based histological analyses showed that oil fraction and ethanolic extracts effectively suppressed the UVB-induced decrease in collagen content. The level of type 1 collagen in the combined oil/ethanol fraction group was enhanced by around 40% compared with the UVB control group. A decrease in UVB-induced intracellular ROS production, and MMP-2 and MMP-9 expression was observed, in comparison with irradiated control cells. Not excluding other compounds, caffeine seems to be a major active compound in the protection against skin UV damage. The results of this study suggest the potential topical application of high caffeine level products as an anti-photoaging agent [48].

2.5. Anti-Cellulite Activity

Cellulite is a skin disorder that affects around 85% to 98% of women. Developed on body areas that contain subcutaneous adipose tissue (upper outer thighs, posterior thighs and buttocks), cellulite is usually known as "orange peel" due to the appearance caused by the herniation of subcutaneous adipose tissue [49]. According to Rawlings [50], cellulite is a multifactorial and complex condition that involves not only subcutaneous fat but also microcirculation and the lymphatic system. Due to their lipolytic effects, caffeine gains increasing interest as a cosmetic ingredient, particularly for gynoid lipodystrophy (commonly known as cellulite). Inhibiting phosphodiesterase in adipocytes and inducing cyclic adenosine monophosphate (cAMP), caffeine could exert lipolytic activity. An increase in cAMP levels stimulates the protein kinase A to phosphorylate, and consequently activates hormone-sensitive lipase. Phosphorylated, hormone-sensitive lipase hydrolyzes triglycerides into di- and monoglycerides, free fatty acids, and glycerol. Caffeine can penetrate the skin barrier and exert its lipolytic activity in adipose tissue. CS contains a significant amount of caffeine (comparable to arabica coffee beans) [51] and, in this context, it can be used to develop anti-cellulite topical products [52]. Indeed, Rodrigues et al. [53] suggests that caffeine extracted from CS can be used in a new possible therapy for cellulite.

2.6. Anti-Hair Loss Activity

Caffeine has been increasingly required for cosmetics production, not only to improve the skin's appearance but also the hair's condition due to its 5α-reductase inhibition activity. This enzyme converts testosterone into the more active dihydrotestosterone (DHT), which is responsible for baldness due to the hair follicles sensitivity to DHT action. Caffeine, by inhibiting 5α-reductase activity, renews the hair growth phase [50]. Indeed, a study of Fischer et al. [54] showed that caffeine concentrations ranging from 0.001% to 0.005% led to in vitro stimulation of human hair follicle growth. The stimulating effects of caffeine on hair growth can also be explained due to the caffeine phosphodiesterase inhibition activity with consequent increase of cAMP intracellular concentration and stimulating cellular metabolism [54]. In addition, caffeine reduces smooth muscle tension near the hair follicle causing an easier delivery of nutrients. Caffeine also arouses capillary vessel microcirculation in the head skin, thereby contributing to nurture hair bulbs. Teichmann et al. [55] and Lademann et al. [56] demonstrated that a 2-min contact of a shampoo with caffeine was sufficient for the formulation to penetrate deeply into the hair follicles and remain there for up to 48 h, even after washing the hair. Otberg et al. [57] showed that there is a quantitative distinction between follicular penetration and interfollicular diffusion of a formulation containing 2.5% caffeine applied to the chest of male Caucasian volunteers aged 26–39 with normal body mass indices. When the follicles remained open, caffeine was detected in blood samples 5 min after topical application. In turn, when the follicles were blocked, caffeine was only detectable after 20 min. The highest values (11.75 ng caffeine/mL) were found 1 h after application, when the follicles were open. The ability of caffeine to penetrate the hair follicles and to stimulate the human hair growth in vitro may have an important clinical impact on the management of androgenetic alopecia, a common problem in men of all ages. However, as far as we know, although CS contains caffeine (0.8–1.25%), no study has focused on its potential to be used in cosmetics for hair loss.

3. Safety and Toxicity

CS seems to be generally safe for use in the cosmetic field. Indeed, Rodrigues and colleagues [17] reported the skin compatibility and safety of CS extracts for topical application. This study, performed with in vitro skin and ocular irritation assays using reconstructed human epidermis (EpiSkin) and a human corneal epithelial model (SkinEthics HCE), respectively, showed that CS extracts were not irritant. Moreover, a histological analysis proved that both model structures were not affected after exposure to the CS extracts. The authors also described an in vivo assay to ensure the safety of a hydroalcoholic extract: a patch test was performed using 20 volunteers during a period of 48 h, after which no skin irritant effects were observed [17]. In addition, other studies [31,32] report that the keratinocytes and fibroblasts viability was not affected due to the inclusion of CS in cream preparations, and no cytotoxic effects were found.

4. Conclusions

CS is a coffee roasting by-product, produced in large amounts every year. Its use can be seen as a gain not only for human health, but also to the environment and industries, answering to the principles of sustainability and circular economy.

Skin aging is a complex process associated to oxidative metabolism and reactive oxygen species (ROS) generation. Silverskin is a promising matrix that can answer to the new fields of cosmetic industry on natural active ingredients that improve health skin appearance, counteract skin aging and related diseases, in an environmentally eco-friendly approach.

One concern can be the presence of OTA (produced by *Aspergillus ochraceus* and *Penicillium verrucosum*) in CS. Nevertheless, good practices of coffee harvesting, storage, and transport (especially related with moisture and temperature exposure) and a rigorous quality control could minimize the presence of this micotoxin.

Due to CS richness in specific bioactive compounds (chlorogenic acids (1–6%), caffeine (0.8–1.25%), and melanoidins (17–23%), among other antioxidants) and confirmed bioactivity in the prevention and/or attenuation of skin aging and related diseases (anti-inflammatory, antimicrobial, anti-cellulite, and anti-hair loss activities, and UV damage protection), CS and its extracts emerge as potential new ingredients for the cosmetics formulation sector.

Acknowledgments: The authors thank the financial support to the project Operação NORTE-01-0145-FEDER-000011—denominada Qualidade e Segurança Alimentar—uma abordagem (nano)tecnológica. This work was also supported by the project UID/QUI/50006/2013—POCI/01/0145/FEDER/007265 with financial support from FCT/MEC through national funds and co-financed by FEDER. S.M.F. Bessada acknowledges the PhD fellowship (SFRH/BD/122754/2016) funded by FCT.

Conflicts of Interest: The authors declare no conflict of interest.

References

1. Alves, R.C.; Rodrigues, F.; Antónia Nunes, M.; Vinha, A.F.; Oliveira, M.B.P.P. State of the art in coffee processing by-products. In *Handbook of Coffee Processing by-Products*, 1st ed.; Galanakis, C.M., Ed.; Academic Press: London, UK, 2017; pp. 1–26.
2. Borrelli, R.C.; Esposito, F.; Napolitano, A.; Ritieni, A.; Fogliano, V. Characterization of a new potential functional ingredient: Coffee silverskin. *J. Agric. Food Chem.* **2004**, *52*, 1338–1343. [CrossRef] [PubMed]
3. Costa, A.S.; Alves, R.C.; Vinha, A.F.; Costa, E.; Costa, C.S.; Nunes, M.A.; Almeida, A.A.; Santos-Silva, A.; Oliveira, M.B.P.P. Nutritional, chemical and antioxidant/pro-oxidant profiles of silverskin, a coffee roasting by-product. *Food Chem.* **2017**. [CrossRef]
4. Costa, A.S.; Alves, R.C.; Vinha, A.F.; Barreira, S.V.; Nunes, M.A.; Cunha, L.M.; Oliveira, M.B.P.P. Optimization of antioxidants extraction from coffee silverskin, a roasting by-product, having in view a sustainable process. *Ind. Crops Prod.* **2014**, *53*, 350–357. [CrossRef]

5.   Carneiro, L.; Silva, J.; Mussatto, S.; Roberto, I.; Teixeira, J. Determination of total carbohydrates content in coffee industry residues. In *Book of Abstracts of the 8th International Meeting of the Portuguese Carbohydrate Group*; GLUPOR: Braga, Portugal, 2009; p. 94.

6.   Napolitano, A.; Fogliano, V.; Tafuri, A.; Ritieni, A. Natural occurrence of ochratoxin A and antioxidant activities of green and roasted coffees and corresponding byproducts. *J. Agric. Food Chem.* **2007**, *55*, 10499–10504. [CrossRef] [PubMed]

7.   Toschi, T.G.; Cardenia, V.; Bonaga, G.; Mandrioli, M.; Rodriguez-Estrada, M.T. Coffee silverskin: Characterization, possible uses, and safety aspects. *J. Agric. Food Chem.* **2014**, *62*, 10836–10844. [CrossRef] [PubMed]

8.   Alves, R.C.; Costa, A.S.; Jerez, M.A.; Casal, S.; Sineiro, J.; Núñez, M.J.; Oliveira, M.B.P.P. Antiradical activity, phenolics profile, and hydroxymethylfurfural in espresso coffee: Influence of technological factors. *J. Agric. Food Chem.* **2010**, *58*, 12221–12229. [CrossRef] [PubMed]

9.   Mesías, M.; Navarro, M.; Martínez-Saez, N.; Ullate, M.; del Castillo, M.; Morales, F. Antiglycative and carbonyl trapping properties of the water soluble fraction of coffee silverskin. *Food Res. Int.* **2014**, *62*, 1120–1126. [CrossRef]

10.   Iriondo-DeHond, A.; Haza, A.I.; Ávalos, A.; del Castillo, M.D.; Morales, P. Validation of coffee silverskin extract as a food ingredient by the analysis of cytotoxicity and genotoxicity. *Food Res. Int.* **2017**, *100*, 791–797. [CrossRef] [PubMed]

11.   European Commission. Commission Regulation (EC) No. 123/2005 of 26 January 2005 amending Regulation (EC) No. 466/2001 as regards ochratoxin A. *Off. J. Eur. Union* **2005**, *L25*, 3–5.

12.   Narita, Y.; Inouye, K. Review on utilization and composition of coffee silverskin. *Food Res. Int.* **2014**, *61*, 16–22. [CrossRef]

13.   Ballesteros, L.F.; Teixeira, J.A.; Mussatto, S.I. Selection of the solvent and extraction conditions for maximum recovery of antioxidant phenolic compounds from coffee silverskin. *Food Bioproc. Technol.* **2014**, *7*, 1322–1332. [CrossRef]

14.   Mussatto, S.I.; Machado, E.M.; Martins, S.; Teixeira, J.A. Production, composition, and application of coffee and its industrial residues. *Food Bioprocess Technol.* **2011**, *4*, 661–672. [CrossRef]

15.   Pourfarzad, A.; Mahdavian-Mehr, H.; Sedaghat, N. Coffee silverskin as a source of dietary fiber in bread-making: Optimization of chemical treatment using response surface methodology. *LWT-Food Sci. Technol.* **2013**, *50*, 599–606. [CrossRef]

16.   Martinez-Saez, N.; Ullate, M.; Martin-Cabrejas, M.A.; Martorell, P.; Genovés, S.; Ramon, D.; del Castillo, M.D. A novel antioxidant beverage for body weight control based on coffee silverskin. *Food Chem.* **2014**, *150*, 227–234. [CrossRef] [PubMed]

17.   Rodrigues, F.; Pereira, C.; Pimentel, F.; Alves, R.; Ferreira, M.; Sarmento, B.; Amaral, M.H.; Oliveira, M.B.P.P. Are coffee silverskin extracts safe for topical use? An in vitro and in vivo approach. *Ind. Crops Prod.* **2015**, *63*, 167–174. [CrossRef]

18.   Rodrigues, F.; Matias, R.; Ferreira, M.; Amaral, M.H.; Oliveira, M.B.P.P. In vitro and in vivo comparative study of cosmetic ingredients coffee silverskin and hyaluronic acid. *Exp. Dermatol.* **2016**, *25*, 572–574. [CrossRef] [PubMed]

19.   Rodrigues, F.; Antónia Nunes, M.; Alves, R.; Oliveira, M. Applications of recovered bioactive compounds in cosmetics and other products. In *Handbook of Coffee Processing by-Products*, 1st ed.; Galanakis, C.M., Ed.; Academic Press: London, UK, 2017; pp. 195–220.

20.   Furusawa, M.; Narita, Y.; Iwai, K.; Fukunaga, T.; Nakagiri, O. Inhibitory effect of a hot water extract of coffee "silverskin" on hyaluronidase. *Biosci. Biotechnol. Biochem.* **2011**, *75*, 1205–1207. [CrossRef] [PubMed]

21.   Rodrigues, F.; Palmeira-de-Oliveira, A.; das Neves, J.; Sarmento, B.; Amaral, M.H.; Oliveira, M.B.P.P. Coffee silverskin: A possible valuable cosmetic ingredient. *Pharm. Biol.* **2015**, *53*, 386–394. [CrossRef] [PubMed]

22.   Moreira, A.S.; Nunes, F.M.; Domingues, M.R.; Coimbra, M.A. Coffee melanoidins: Structures, mechanisms of formation and potential health impacts. *Food Funct.* **2012**, *3*, 903–915. [CrossRef] [PubMed]

23.   Lephart, E.D. Skin aging and oxidative stress: Equol's anti-aging effects via biochemical and molecular mechanisms. *Ageing Res. Rev.* **2016**, *31*, 36–54. [CrossRef] [PubMed]

24.   Berthon, J.-Y.; Nachat-Kappes, R.; Bey, M.; Cadoret, J.-P.; Renimel, I.; Filaire, E. Marine algae as attractive source to skin care. *Free Rad. Res.* **2017**, *51*, 555–567. [CrossRef] [PubMed]

25. Pisoschi, A.M.; Pop, A. The role of antioxidants in the chemistry of oxidative stress: A review. *Eur. J. Med. Chem.* **2015**, *97*, 55–74. [CrossRef] [PubMed]

26. Bessada, S.M.; Barreira, J.C.; Oliveira, M.B.P.P. Asteraceae species with most prominent bioactivity and their potential applications: A review. *Ind. Crops Prod.* **2015**, *76*, 604–615. [CrossRef]

27. Del Castillo, M.; Fernandez-Gomez, B.; Martinez Saez, N.; Iriondo De Hond, A.; Martirosyan, D.; Mesa, M. Coffee silverskin extract for aging and chronic diseases. In *Functional Foods for Chronic Diseases*, 1st ed.; Martirosyan, D.M., Ed.; CreateSpace Independent Publishing Platform: Colorado, TX, USA, 2016; pp. 386–409.

28. Tobin, D.J. Introduction to skin aging. *J. Tissue Viability* **2017**, *26*, 37–46. [CrossRef] [PubMed]

29. Paur, I.; Balstad, T.R.; Blomhoff, R. Degree of roasting is the main determinant of the effects of coffee on NF-κb and epre. *Free Radic. Biol. Med.* **2010**, *48*, 1218–1227. [CrossRef] [PubMed]

30. Sahasrabudhe, A.; Deodhar, M. Anti-hyaluronidase, anti-elastase activity of *Garcinia indica*. *Int. J. Bot.* **2010**, *6*, 1–10. [CrossRef]

31. Rodrigues, F.; Gaspar, C.; Palmeira-de-Oliveira, A.; Sarmento, B.; Helena Amaral, M.; Oliveira, M.B.P.P. Application of coffee silverskin in cosmetic formulations: Physical/antioxidant stability studies and cytotoxicity effects. *Drug Dev. Ind. Pharm.* **2016**, *42*, 99–106. [CrossRef] [PubMed]

32. Rodrigues, F.; Sarmento, B.; Amaral, M.H.; Oliveira, M.B.P.P. Exploring the antioxidant potentiality of two food by-products into a topical cream: Stability, in vitro and in vivo evaluation. *Drug Dev. Ind. Pharm.* **2016**, *42*, 880–889. [CrossRef] [PubMed]

33. Iriondo-DeHond, A.; Martorell, P.; Genovés, S.; Ramón, D.; Stamatakis, K.; Fresno, M.; Molina, A.; del Castillo, M.D. Coffee silverskin extract protects against accelerated aging caused by oxidative agents. *Molecules* **2016**, *21*, 721. [CrossRef] [PubMed]

34. Martinez, R.M.; Pinho-Ribeiro, F.A.; Steffen, V.S.; Silva, T.C.; Caviglione, C.V.; Bottura, C.; Fonseca, M.J.; Vicentini, F.T.; Vignoli, J.A.; Baracat, M.M. Topical formulation containing naringenin: Efficacy against ultraviolet b irradiation-induced skin inflammation and oxidative stress in mice. *PLoS ONE* **2016**, *11*, e0146296. [CrossRef] [PubMed]

35. Menezes, A.C.; Campos, P.M.; Euletério, C.; Simões, S.; Praça, F.S.G.; Bentley, M.V.L.B.; Ascenso, A. Development and characterization of novel 1-(1-naphthyl) piperazine-loaded lipid vesicles for prevention of uv-induced skin inflammation. *Eur. J. Pharm. Biopharm.* **2016**, *104*, 101–109. [CrossRef] [PubMed]

36. Shin, H.S.; Satsu, H.; Bae, M.-J.; Zhao, Z.; Ogiwara, H.; Totsuka, M.; Shimizu, M. Anti-inflammatory effect of chlorogenic acid on the IL-8 production in Caco-2 cells and the dextran sulphate sodium-induced colitis symptoms in c57bl/6 mice. *Food Chem.* **2015**, *168*, 167–175. [CrossRef] [PubMed]

37. Hwang, J.-H.; Koh, E.-J.; Lee, Y.-J.; Chio, J.; Song, J.-H.; Seo, Y.-J.; Lee, B.-Y. Anti-inflammatory effect of caffeine by regulating NF-κb activation in murine macrophage. *FASEB J.* **2016**, *30*, lb256. Available online: http://www.fasebj.org/content/30/1_Supplement/lb256.short (accessed on 3 January 2018).

38. Vitaglione, P.; Morisco, F.; Mazzone, G.; Amoruso, D.C.; Ribecco, M.T.; Romano, A.; Fogliano, V.; Caporaso, N.; D'argenio, G. Coffee reduces liver damage in a rat model of steatohepatitis: The underlying mechanisms and the role of polyphenols and melanoidins. *Hepatology* **2010**, *52*, 1652–1661. [CrossRef] [PubMed]

39. Jiménez-Zamora, A.; Pastoriza, S.; Rufián-Henares, J.A. Revalorization of coffee by-products. Prebiotic, antimicrobial and antioxidant properties. *LWT-Food Sci. Technol.* **2015**, *61*, 12–18. [CrossRef]

40. Rufian-Henares, J.A.; de la Cueva, S.P. Antimicrobial activity of coffee melanoidins: A study of their metal-chelating properties. *J. Agric. Food Chem.* **2009**, *57*, 432–438. [CrossRef] [PubMed]

41. Antonio, A.G.; Moraes, R.S.; Perrone, D.; Maia, L.C.; Santos, K.R.N.; Iório, N.L.; Farah, A. Species, roasting degree and decaffeination influence the antibacterial activity of coffee against streptococcus mutans. *Food Chem.* **2010**, *118*, 782–788. [CrossRef]

42. Cardenia, V.; Rodriguez-Estrada, M.T.; Boselli, E.; Lercker, G. Cholesterol photosensitized oxidation in food and biological systems. *Biochimie.* **2013**, *95*, 473–481. [CrossRef] [PubMed]

43. Kitagawa, S.; Yoshii, K.; Morita, S.-Y.; Teraoka, R. Efficient topical delivery of chlorogenic acid by an oil-in-water microemulsion to protect skin against UV-induced damage. *Chem. Pharm. Bull.* **2011**, *59*, 793–796. [CrossRef] [PubMed]

44. Quan, T.; Qin, Z.; Xia, W.; Shao, Y.; Voorhees, J.J.; Fisher, G.J. Matrix-degrading metalloproteinases in photoaging. *J. Investig. Dermatol. Symp. Proc.* **2009**, *14*, 20–24. [CrossRef] [PubMed]

45. Cho, Y.-H.; Bahuguna, A.; Kim, H.-H.; Kim, D.-I.; Kim, H.-J.; Yu, J.-M.; Jung, H.-G.; Jang, J.-Y.; Kwak, J.-H.; Park, G.-H. Potential effect of compounds isolated from *Coffea arabica* against UV-B induced skin damage by protecting fibroblast cells. *J. Photochem. Photobiol. B Biol.* **2017**, *174*, 323–332. [CrossRef] [PubMed]

46. Lu, Y.-P.; Lou, Y.-R.; Xie, J.-G.; Peng, Q.-Y.; Liao, J.; Yang, C.S.; Huang, M.-T.; Conney, A.H. Topical applications of caffeine or (−)-epigallocatechin gallate (EGCG) inhibit carcinogenesis and selectively increase apoptosis in UVB-induced skin tumors in mice. *Proc. Natl. Acad. Sci. USA* **2002**, *99*, 12455–12460. [CrossRef] [PubMed]

47. Koo, S.W.; Hirakawa, S.; Fujii, S.; Kawasumi, M.; Nghiem, P. Protection from photodamage by topical application of caffeine after ultraviolet irradiation. *Br. J. Dermatol.* **2007**, *156*, 957–964. [CrossRef] [PubMed]

48. Choi, H.-S.; Park, E.D.; Park, Y.; Han, S.H.; Hong, K.B.; Suh, H.J. Topical application of spent coffee ground extracts protects skin from ultraviolet B-induced photoaging in hairless mice. *Photochem. Photobiol. Sci.* **2016**, *15*, 779–790. [CrossRef] [PubMed]

49. Alizadeh, Z.; Halabchi, F.; Mazaheri, R.; Abolhasani, M.; Tabesh, M. Review of the mechanisms and effects of noninvasive body contouring devices on cellulite and subcutaneous fat. *Int. J. Endocrinol. Metab.* **2016**, *14*, e36727. [CrossRef] [PubMed]

50. Rawlings, A. Cellulite and its treatment. *Int. J. Cosmet. Sci.* **2006**, *28*, 175–190. [CrossRef] [PubMed]

51. Alves, R.C.; Oliveira, M.B.P.P.; Casal, S. Coffee authenticty. In *Current Topics on Food Authetication*, 1st ed.; Oliveira, M.B.P.P., Mafra, I., Amaral, J.S., Eds.; Transworld Research Network: Kerala, India, 2017; pp. 57–72.

52. Herman, A.; Herman, A. Caffeine's mechanisms of action and its cosmetic use. *Skin Pharmacol. Physiol.* **2013**, *26*, 8–14. [CrossRef] [PubMed]

53. Rodrigues, F.; Alves, A.C.; Nunes, C.; Sarmento, B.; Amaral, M.H.; Reis, S.; Oliveira, M.B.P.P. Permeation of topically applied caffeine from a food by-product in cosmetic formulations: Is nanoscale in vitro approach an option? *Int. J. Pharm.* **2016**, *513*, 496–503. [CrossRef] [PubMed]

54. Fischer, T.; Hipler, U.; Elsner, P. Effect of caffeine and testosterone on the proliferation of human hair follicles in vitro. *Int. J. Dermatol.* **2007**, *46*, 27–35. [CrossRef] [PubMed]

55. Teichmann, A.; Richter, H.; Knorr, F.; Antoniou, C.; Sterry, W.; Lademann, J. Investigation of the penetration and storage of a shampoo formulation containing caffeine into the hair follicles by in vivo laser scanning microscopy. *Laser Phys. Lett.* **2007**, *4*, 464–468. [CrossRef]

56. Lademann, J.; Richter, H.; Schanzer, S.; Klenk, A.; Sterry, W.; Patzelt, A. Analysis of the penetration of a caffeine containing shampoo into the hair follicles by in vivo laser scanning microscopy. *Laser Phys.* **2010**, *20*, 551–556. [CrossRef]

57. Otberg, N.; Patzelt, A.; Rasulev, U.; Hagemeister, T.; Linscheid, M.; Sinkgraven, R.; Sterry, W.; Lademann, J. The role of hair follicles in the percutaneous absorption of caffeine. *Br. J. Clin. Pharm.* **2008**, *65*, 488–492. [CrossRef] [PubMed]

Article

Personal-Care Products Formulated with Natural Antioxidant Extracts

Maria Luisa Soto [1,2,*], María Parada [1], Elena Falqué [3] and Herminia Domínguez [1,*]

[1] Departamento de Enxeñería Química, Universidade de Vigo (Campus Ourense), Edificio Politécnico, As Lagoas, Ourense 32004, Spain; mariapc@uvigo.es

[2] Departamento de Imaxe Persoal, IES Lauro Olmo, O Barco de Valdeorras, Ourense 32350, Spain

[3] Departamento de Química Analítica, Universidade de Vigo (Campus Ourense), Edificio Politécnico, As Lagoas, Ourense 32004, Spain; efalque@uvigo.es

* Corresponding author: sotoalvarez@edu.xunta.es (M.L.S.); herminia@uvigo.es (H.D.)

Received: 5 December 2017; Accepted: 16 January 2018; Published: 18 January 2018

Abstract: The objective of this study was to evaluate the potential use of some vegetal raw materials in personal-care products. Four ethanolic extracts (grape pomace, *Pinus pinaster* wood chips, *Acacia dealbata* flowers, and *Lentinus edodes*) were prepared and total phenolics, monomeric sugars, and antioxidant capacity were determined on alcoholic extracts. Six of the most important groups of cosmetics products (hand cream, body oil, shampoo, clay mask, body exfoliating cream, and skin cleanser) were formulated. Participants evaluated some sensory attributes and overall acceptance by a 10-point scale; the results showed differences among age-intervals, but not between males and females. The results confirmed that all extracts presented characteristics appropriate for their use in cosmetic formulations and their good acceptability by consumers into all cosmetic products. Texture/appearance, spreadability, and skin feeling are important attributes among consumer expectations, but odor and color were the primary drivers and helped differentiate the natural extracts added into all personal-care products.

Keywords: grape pomace; pine wood; acacia flowers; shiitake; cosmetics; sensory acceptability

1. Introduction

Consumers are increasingly demanding natural ingredients and additives in cosmetic products, as well as the replacement of synthetic compounds with possible negative effects on health and the environment [1].

Antioxidants are preservatives with the function of preventing lipid oxidation of the product. They can act following different mechanisms, i.e., reducing agents, oxygen scavengers, synergistic agents, and chelating agents. More recently, antioxidant-based products have been proposed to protect the skin [2,3].

Among natural compounds with antioxidant properties, phenolics are the most studied. Natural phenolics—including benzoic acids, cinnamic acids, and flavonoids—are widely distributed in renewable and abundant sources, such as agricultural, food, and forest products and by-products. The utilization of these alternative low cost sources is desirable for the integral valorization of vegetal raw materials and could benefit the economy of the process and the cost of the products. The potential of selected natural extracts obtained from underutilized and residual vegetal biomass processed with food-grade green solvents as additives in cosmetic products was previously reported. The extracts were safe for topical use and enhanced the oxidative stability of model oil-in-water emulsions [4].

Phenolic compounds present a wide variety of activities of interest in cosmetics, such as antioxidant, antimicrobial, anti-inflammatory, or anti-aging. Formulations enriched in phenolic antioxidants are increasingly used in anti-aging cosmetics as a defense strategy against reactive oxygen species (ROS) [5].

In addition, natural phenolic compounds can permeate through the skin barrier, in particular the stratum corneum [6].

Cosmetic products need to be effective and stable [7], but also the acceptance by the consumer needs to be confirmed. Equally important are the favorable benefits on skin health and the desirable sensory attributes [8], because the incorporation of natural extracts could confer undesirable characteristics and strong colors or aromas, that would limit the acceptance of the product.

Sensory analysis can discriminate the characteristics influencing consumer acceptance and to indicate how they are perceived and, consequently, guide success in the development of new cosmetic products. According to Elezoviü et al. [9], the characteristics influencing consumer acceptance are based, first, on its packaging, and then on its smell, appearance, and texture (touch and feel). Therefore, after the development of a formulation, researchers and cosmetic companies should carry out a sensory evaluation with trained or consumer panels. Sensory analysis represents a valuable tool, but it is financially expensive and time-consuming. For this reason, some papers have recently appeared studying the application of instrumental analysis, mainly through rheological measurements, to detect changes of entry ingredients [10,11]. Cosmetic properties, such as the optimal mechanical properties (firmness), adequate rheological behavior, and appropriate adhesion, could be measured by instrumental analysis. Other attributes—including, appearance, odor, residual greasiness after application, or the sensation produced by the cosmetic application—play an important role in the acceptability of cosmetic products by consumers. They are subjective and, consequently sensory evaluation methods should be applied [12,13].

The objective of this work was to formulate six cosmetic products with ethanolic extracts from four vegetal raw materials which provide different types of aromatic families: flower, fruity, wood, and mushroom/earthy. The reducing power and radical scavenging capacity of the extracts were characterized and the sensory evaluation of the final personal-care products was assessed.

2. Materials and methods

2.1. Materials

Grape pomace was provided by Destilería Galicia (O Barco de Valdeorras, Ourense, Spain). Pine (*Pinus pinaster*) wood chips, kindly provided by FINSA Orember (Ourense, Spain), were air dried and milled under 1 mm. *Acacia dealbata* flowers were collected in forest areas in the surrounding of Ourense (Spain) in Winter 2014 and freshly processed. Shiitake (*Lentinus edodes*) was purchased in local markets and was freeze dried and ground before processing.

2.2. Extracts

Pressed distilled grape pomace was extracted with water at a liquid to solid ratio 15 w/w, at 50 °C in an orbital shaker at 175 rpm overnight. The solid and liquid phases were separated by filtration and the liquid phase was contacted with non-ionic polymeric resins (Sepabeads SP700, Resindion S.R.L., Mitsubishi Chemical Corp., Milan, Italy). Before use, resins were rinsed with deionized water at a liquid-to-solid ratio of 5 (w/w). Desorption was carried out with 96% ethanol at a solvent to resin ratio 3 (mL/g) in an orbital shaker at 175 rpm and 50 °C. The resin was regenerated in 1 M NaOH overnight and further washed with deionized water [14].

Ground *Pinus pinaster* wood samples, *Acacia dealbata* flowers, and ground freeze-dried *Lentinus edodes* samples were contacted with 96% ethanol in sealed Erlenmeyer flasks at 50 °C in an orbital shaker at 175 rpm overnight.

2.3. Cosmetic Formulations

Six cosmetic model products were formulated with conventional ingredients and with the extracts from the selected sources. The cosmetics prepared were: hand cream (HC), body oil (BO), shampoo (S), clay mask (CM), body exfoliating (BE), and a skin cleanser (SC). The ethanolic extracts—grape pomace extract (GPE), pine wood extract (PWE), acacia flowers extract (AFE), and shiitake extract (SE)—were added to cosmetics dissolved in a minimum amount of ethanol. These extracts have both the function of antioxidants and additives (colorants and perfumes). Control samples without extracts were also prepared.

Hand cream (HC) contained: paraffinun liquidum (30 g), lanolin (30 g), Kathon CG (0.2 g), and extract (five drops). Paraffin was slowly melted in a water bath (50 °C) and stirred to obtain an homogeneous mixture, which was neutralized with triethanolamine (if required). The extracts were added to the cold mixture.

Body oil (BO) was prepared with: almond oil (20 mL), *Glycine soya* oil (20 mL), *Ricinus communis* oil (10 mL). The oils were mixed with gentle stirring in a water bath at 40 °C, and once cooled, the extract was added.

Shampoo (S) was prepared with the following ingredients: sodium laureth sulfate (45 wt %), diethanolamine (3 wt %), Kathon CG (0.02 wt %), citric acid (0.01 wt %), extract (five drops) and distilled water. Approximately half of the water volume was mixed with the detergent and the other half was used to dissolve diethanolamine, citric acid and Kathon. Both solutions were mixed with intense stirring before adding the extract.

Clay mask (CM) was formulated with the following components: sodium laureth sulfate (0.1 g), kaolin (35 g), bentonite (5 g), cetyl alcohol (2 g), glycerin (10 g), Kathon CG (0.2 g), extract (five drops), and distilled water. Water was incorporated to bentonite and was allowed to stand 24 h until gelification. Cetyl alcohol was melted in a water bath. The detergent, glycerin, and the antimicrobial agent were added to the bentonite mixture, which was then heated at 40 °C. Kaolin was also added at this temperature, stirring to avoid lumps. The extract was added to the cooled mixture.

Body exfoliating salt scrub (BE) was composed of: sodium chloride (150 g), almond oil and extract (five drops). The oil was used in an amount needed to moist the salt, then the extract was added and the mixture was stirred until homogenization.

Skin cleanser (SC) was prepared with: cetyl alcohol (10.5 mL), liquid paraffin (30 mL), distilled water (258 mL), triethanolamine (1.5 mL), Kathon CG (0.6 mL), and extract (five drops). Cetyl alcohol was melted in a water bath at 70 °C and mixed with paraffin under mild stirring. Triethanolamine, the antimicrobial agent, and distilled water were mixed with continuous stirring at the previously indicated temperature. The aqueous phase was dropped on the oily phase until it coole and then the extract was added.

All cosmetic ingredients were purchased from Guinama (Valencia, Spain).

2.4. Analytical Methods

The extracts were characterized for phenolic and sugar content and for antioxidant activity. The personal care products were characterized for sensorial properties. Total phenolic content was colorimetrically determined using the Folin–Ciocalteu reagent (Sigma-Aldrich, St. Louis, MO, USA) and expressed as gallic acid (Sigma-Aldrich, St. Louis, MO, USA) equivalents. All analyses were performed at least in triplicate and are reported on a dry matter basis. Ash content was gravimetrically determined.

The monomeric sugars: glucose, xylose, mannose, and galactose, estimated from the concentrations of monosaccharides in samples previously hydrolyzed with 4% sulfuric acid at 121 °C for 20 min, were assayed by HPLC in a 1100 series Hewlett-Packard chromatograph fitted with a refractive index RI detector and a 300×7.8 mm Aminex HPX-87H column (BioRad, Hercules, CA, USA) operating at 60 °C (mobile phase: 0.003 M H_2SO_4, flow rate: 0.6 mL/min) [15].

2.5. Antioxidant Activity

Ferric reducing antioxidant power (FRAP): The reagent, also used as blank, was prepared with 25 mL of 300 mmol/L acetate buffer (pH 3.6), 2.5 mL of a 10 mmol/L 2,4,6-tripyridyl-s-triazine (TPTZ) solution in 40 mmol/L HCl and 20 mmol/L $FeCl_3 \cdot 6H_2O$ in distilled water. Samples (100 µL) were mixed with 3 mL reagent, and the absorbance was monitored at 593 nm and compared with standard aqueous solutions of ascorbic acid.

ABTS (ABTS$^{\bullet+}$, 2,2′-azinobis (3-ethyl-benzothiazoline-6-sulfonate)): ABTS was produced by reacting 7 mM ABTS stock solution with 2.45 mM potassium persulfate. The ABTS$^{\bullet+}$ solution was diluted with PBS (pH = 7.4) to an absorbance of 0.70 at 734 nm and equilibrated at 30 °C. After addition of 1.0 mL of diluted ABTS$^{\bullet+}$ solution to 10 μL of extract or Trolox in ethanol or PBS the absorbance was read at 734 nm and 30 °C. The percentage was calculated as a function of the concentration of extracts and Trolox, and expressed as TEAC (Trolox equivalent antioxidant capacity).

2.6. Sensory Analysis

The sensory panel consisted of 26 female and 29 male assessors (18–45 years old) recruited from a pool of students and staff of IES Lauro Olmo (O Barco de Valdeorras, Ourense, Spain), without previous experience in sensory analysis. The six cosmetic samples (about 10 mL) were served in transparent glass containers encoded with three-digit random numbers. Mineral water and paper towels were provided for skin rinsing between samples.

Panelists were asked to fill out a questionnaire evaluating the intensity of the sensory properties by using a 0 to 10 scale (where 0 represents 'none' and 10 'extremely strong') in two sessions along one month. The list contained 15 descriptors or attributes typically used to characterize the skincare products: six appearance attributes (gloss, color, odor intensity, odor preference, firmness/consistency, creaminess/texture/appearance) and eight skin parameters (spreadability, penetration, softness, skin odor intensity, skin odor preference, skin odor persistence, skin gloss, and skin feel). Finally, participants were asked to rate their global appreciation of the product on a 10-point scale to report which extract they preferred in each formulation. Assessors valuated the appearance attributes during the first session; the skin parameters and the comparison of the six personal-care products to the global score were recorded in the second session.

2.7. Statistical Analysis

Intensity values from sensorial data were analyzed by a two-factor (extract, cosmetic) analysis of variance (ANOVA) test using Excel software. A Fisher LSD post hoc test ($p < 0.05$) was used to test the significance of the relative mean differences among the samples. Differences among extracts and sample formulations were obtained from preference and descriptive data evaluated by means of principal component analysis (PCA) using the program Statistica 8.0 (Statsoft Inc., 2004, Tulsa, OK, USA).

3. Results and Discussion

3.1. Antioxidant Properties

The chemical, physical, and antioxidant properties of the selected extracts are shown in Table 1. The extraction yields were 3.1 g extract/100 g grape pomace, 2.8 g extract/100 g pine wood, 16.9 g extract/100 g acacia flower, and 12.8 g extract/100 g shiitake. The extraction yield attained with the different sources was low, under 20% of the initial material. Therefore, the utilization of the solid for energy or bioconversion purposes could be proposed for further valorization.

The phenolic content in the extracts was higher for the grape pomace and acacia flower (18 wt %), and very low for the mushroom, which contained sugars and polyols (threalose, mannitol, and arabitol). The increased phenolic content of the extracts led to increased ABTS radical scavenging capacity and reducing capacity, except for the grape pomace extract. The most active radical scavengers were the acacia flower extract (AFE), followed by the grape pomace extract (GPE) and the pine wood extract (PWE), with 50–70% of the activity of Trolox. The highest reducing power was also found for AFE.

Table 1. Characterization of extracts obtained from renewable underutilized sources: grape pomace extract (GPE), pine wood extract (PWE), acacia flower extracts (AFE), and shiitake extract (SE).

	Extract			
	GPE	PWE	AFE	SE
Extraction yield (%)	3.2	2.8	16.9	12.8
Physicochemical properties				
Composition (g/100 g extract)				
Phenolic content (gallic acid equiv.)	18.3	9.6	17.9	0.9
Total N	0.86	<0.10	1.24	2.06
Total sugars	n.d.	Glucose (2.64) Xylose (2.48) Arabinose (2.19)	Glucose (9.41) Xylose (3.38)	Glucose (0.99) Trehalose (14.52) Mannitol (10.02) Arabitol (20.04)
In vitro antioxidant properties				
TEAC (g Trolox/g extract)	0.52	0.32	0.70	0.01
FRAP (g AA/g extract)	0.01	0.10	0.17	<0.01

AA: Ascorbic acid; TEAC: Trolox Equivalent Antioxidant Capacity; FRAP: Ferric Reducing Antioxidant Power; n.d.: non detected; $\text{TEAC}_{\text{gallic acid}}$: 4.93 g Trolox/g extract; $\text{FRAP}_{\text{gallic acid}}$: 0.17 g AA/g extract.

3.2. Sensorial Evaluation

Many natural products, botanicals, or waste materials—derived from agricultural products, foods and beverages—can be used in cosmetics products [16]. Sensory test was carried out to evaluate the possibility of using natural extracts as ingredients of some cosmetic preparations and their acceptability by consumers. Independently of the antioxidant activity, the ethanolic extracts were added to provide their odoriferous characteristics to the different cosmetic formulations and to evaluate their acceptance by the consumers. Likewise, the color differed with raw material too and the volunteers scored also this difference. Personal-care products from acacia flowers extract (AFE) always showed an intense yellow color (Figure 1).

Figure 1. Personal-care products elaborated without extract or with grape pomace, pine wood, acacia flower, and shiitake ethanolic extracts.

Sensory analysis was performed to determine the preference for the natural extracts, because the composition of the individual formulations was the same, except in this ingredient. A control sample which did not contain any extract from the studied vegetal raw materials was used for reference. The most preferred and valued attributes in all cosmetics were spreadability, softness, consistence/texture, and skin feel, but the ANOVA results showed that these attributes in each cosmetic product were comparable and the use of the different extracts caused only a significant effect ($p < 0.05$) on two parameters: color and odor.

Participants from the two genres accomplished the sensory analysis (26 women and 29 men), and, as can be seen in Figure 2, no great differences between them were found. Only four samples demonstrated significant differences: women evaluated better than men the acacia flower extract when it was included in the hand cream and exfoliating preparations, and the control-exfoliating and the shampoo with grape pomace extract obtained better scores with men than with women. Female participants valued body oil and clay masks better than males, and, on the contrary, men valued the shampoo more. Hand cream elaborated with acacia flowers or shiitake attained the best scores, with 7.20 and 7.04 points, respectively.

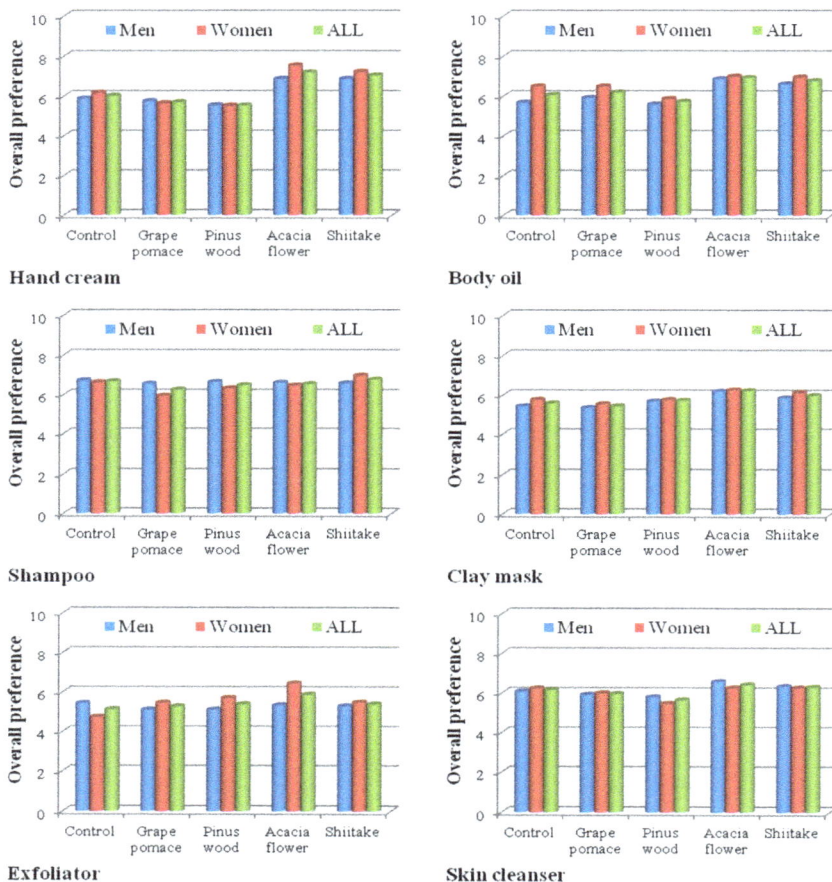

Figure 2. Overall appreciation for tested personal-care products according to the volunteers' gender.

From the four assayed raw materials, the least preferred extracts in all cosmetics were red grape pomace and pinus wood ethanolic extracts. The average overall preference showed that these extracts were best valued as aromatic additives in the shampoo, and the worst score was attained by the exfoliating body. The cosmetic most valued with the acacia or shiitake extracts were the hand creams, and the worst was the body exfoliating cream. Control samples (without any added extracts) achieved similar scores than when GPE and PWE were added into all personal-care products, except in shampoo, where it was obtained the best punctuation along with the shiitake extract.

The participants were also divided into three age-segments: <20, 20–30, and 30–45 years, with 20, 19, and 16 persons, respectively. This last group perceived acacia and shiitake extracts with the significant highest values for all formulations, revealing the influence of color and odor of these natural extracts (Figure 3). In opposition, the participants under 20 years of age preferred the extracts obtained from pinus wood and grape pomace or the control sample (without any extract), except in hand cream. These two extracts received lower score in the group of older participants. Consumers in the 20–30 years old segment evaluated samples with the highest overall preference values in all personal-care products, and the preferred extracts were acacia flowers and shiitake extracts.

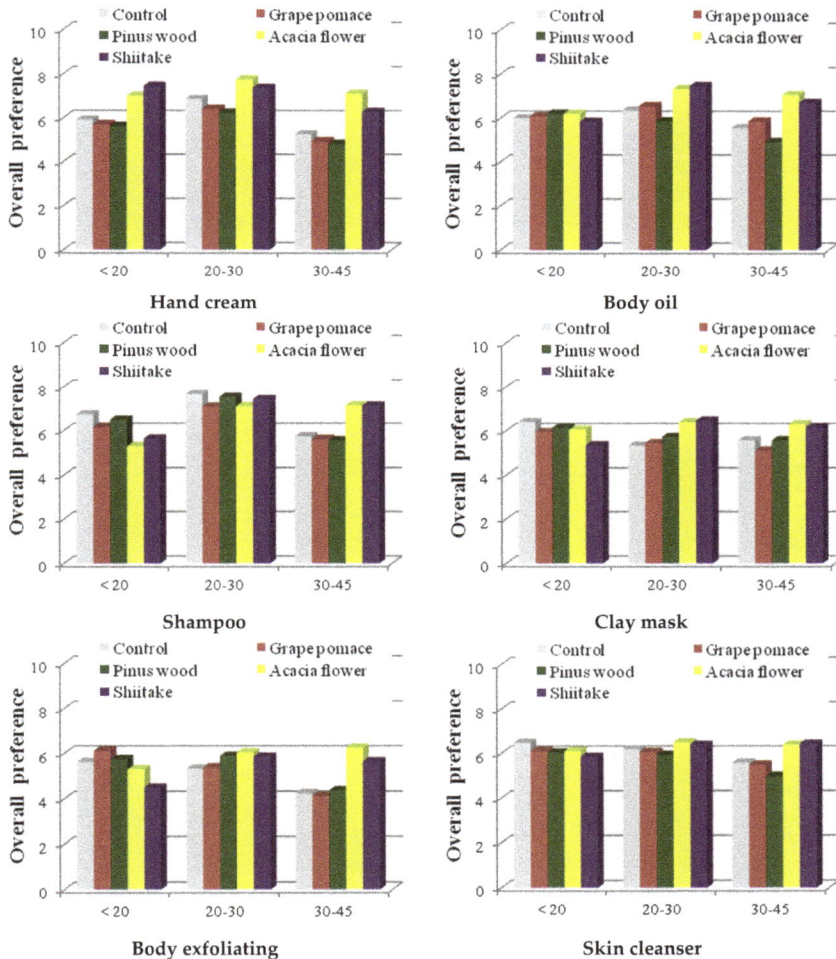

Figure 3. Overall appreciation for tested personal-care products according to the volunteers' age-interval.

The sensory characterization confirms which properties mostly influence consumer acceptance. All volunteers considered that the different tested personal-care products have a good spreadability, softness, and good skin penetration ability and posterior skin feeling; but also a pleasant color and fragrance. These results have shown that a variety of plant materials can be used as additives in cosmetic products to supply color and aroma [17–19].

Principal component analysis (Figure 4) was performed to know how consumer acceptance is based on sensory attributes. The interrelationships among the extracts from natural raw materials used in personal care-products and sensory descriptors showed that the first two factors explained 88.3% of the total variance among the extracts. The first component accounted for 70.5% of the data variability and the second for 17.8%. In the PCA the extract samples are clearly grouped into two clusters. Cluster 1, located in the upper half, was associated with the cosmetics elaborated from acacia flowers and shiitake extracts, which had the greatest acceptance. Cluster 2, in the positive part of Factor 1 and in the negative part of Factor 2, was characterized by the other two extracts (grape pomace and pine wood) and the control (without added extract). No differences between different cosmetic formulations were found.

All products were well-tolerated because any visible skin irritation or erythema was observed. Besides expected appearance, spreadability, softness, and skin feeling, the color and odor also play an important role on overall preference and, consequently, on purchase intent. According to the obtained results, fragrance and color were two important attributes for consumer preference and they are essential additives to make personal-care products, even, cosmetic companies use colors in packaging design to communicate the properties of their fragrances [20]. Between the four assayed vegetal extracts, the floral aroma and yellow color provided by the acacia flower extract were evaluated higher by all consumers, independently of genre or age; likewise, this ethanolic extract also presented the highest in vitro antioxidant activity.

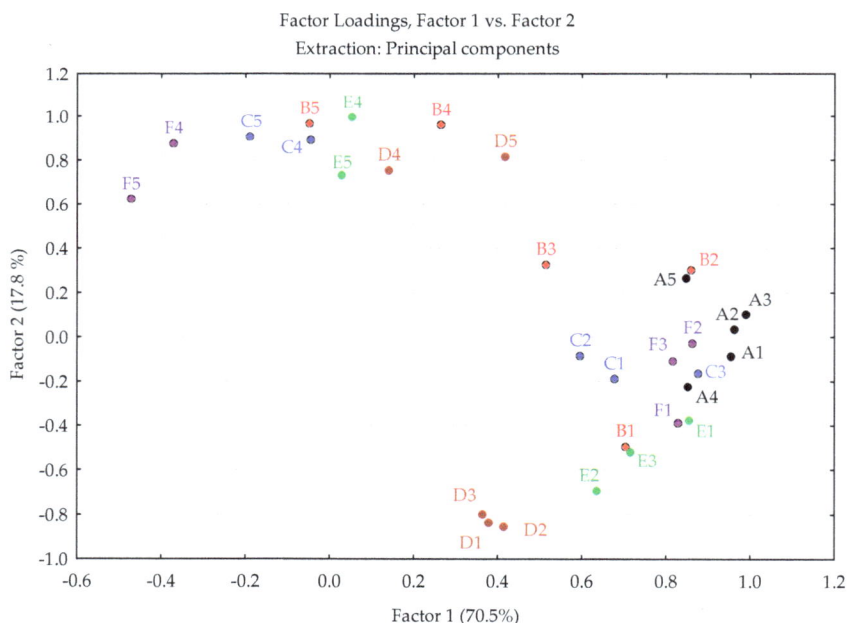

Figure 4. Principal component analysis bi-plot of cosmetic formulations. A–F indicate hand cream (HC), body oil (BO), shampoo (S), clay mask (CM), body exfoliating cream (BE), and a skin cleanser formulation (SC) respectively. 1–5 indicate control (without extract) grape pomace extract (GPE), pine wood extract (PWE), acacia flower extracts (AFE), and shiitake extract (SE), respectively.

4. Conclusions

Consumers are increasingly rejecting synthetic chemicals in beauty and cosmetic products and demand natural products. In this study, ethanolic extracts from different raw materials were added to

Cosmetics **2018**, *5*, 13

some personal-care products and sensory analysis was carried out to evaluate some attributes such as color, aroma, and texture, and the overall acceptability by consumers. Results have shown that the ethanolic extracts obtained from a flower, a mushroom, a tree, or agricultural waste (grape pomace) can be valorized and they have a potential application as an ingredient for cosmetic formulations.

Acknowledgments: The authors are grateful to the Spanish Ministry of Education and Science (Research Project reference CTM2015-68503-R) and to Xunta de Galicia (IN852A 2013/63-0) for the financial support of this work. M. Parada thanks Xunta de Galicia (INBIOMED) for funding. Both projects were partially funded by the FEDER Program of the European Union ("Unha maneira de facer Europa"). The authors thank to Miguel Estévez for his technical support.

Author Contributions: Maria Luisa Soto and Herminia Domínguez conceived and designed the experiments; Maria Luisa Soto and Elena Falqué designed the sensory analysis; Maria Luisa Soto and María P. Casas performed the experiments; Elena Falqué and Herminia Domínguez analyzed the data; Herminia Domínguez contributed reagents/materials/analysis tools; Elena Falqué and Herminia Domínguez wrote the paper.

Conflicts of Interest: The authors declare no conflict of interest.

References

1. Chermahini, S.H.; Majid, F.A.A.; Sarmidi, M.R. Cosmeceutical value of herbal extracts as natural ingredients and novel technologies in anti-aging. *J. Med. Plants Res.* **2011**, *5*, 3074–3077.
2. Andreassi, M.; Andreassi, L. Antioxidants in dermocosmetology: From the laboratory to clinical application. *J. Cosmet. Dermatol.* **2004**, *2*, 153–160. [CrossRef] [PubMed]
3. Buenger, J.; Ackermann, H.; Jentzsch, A.; Mehling, A.; Pfitzner, I.; Reiffen, K.A.; Schroeder, K.R.; Wollenweber, U. An interlaboratory comparison of methods used to assess antioxidant potentials. *Int. J. Cosmet. Sci.* **2006**, *28*, 135–146. [CrossRef] [PubMed]
4. Balboa, E.M.; Soto, M.L.; Nogueira, D.R.; González-López, N.; Conde, E.; Moure, A.; Vinardell, M.P.; Mitjans, M.; Domínguez, H. Potential of antioxidant extracts produced by aqueous processing of renewable resources for the formulation of cosmetics. *Ind. Crops Prod.* **2014**, *58*, 104–110. [CrossRef]
5. Pouillot, A.; Polla, L.L.; Tacchini, P.; Neequaye, A.; Polla, A.; Polla, B. Natural antioxidants and their effects on the skin. Chapter 13. In *Formulating, Packaging, and Marketing of Natural Cosmetic Products*, 1st ed.; John Wiley & Sons Press: Hoboken, NJ, USA, 2011.
6. Zillich, O.V.; Schweiggert-Weisz, U.; Hasenkopf, K.; Eisner, P.; Kerscher, M. Release and in vitro skin permeation of polyphenols from cosmetic emulsions. *Int. J. Cosmet. Sci.* **2013**, *35*, 491–501. [CrossRef] [PubMed]
7. Guidelines on Stability Testing of Cosmetic Products. Available online: https://www.cosmeticseurope. eu/files/3714/6407/8024/Guidelines_for_the_Safety_Assessment_of_a_Cosmetic_Product_-_2004.pdf (accessed on 1 December 2017).
8. Guest, S.; McGlone, F.; Hopkinson, A.; Schendel, Z.A.; Blot, K.; Essick, G. Perceptual and sensory-functional consequences of skin care products. *J. Cosm. Dermatol. Sci. Appl.* **2013**, *3*, 66–78. [CrossRef]
9. Elezoviü, A.; Hadžiabdiü, J.; Rahiü, O.; Vraniü, E. Measuring the feeling: Correlations of sensorial to instrumental analyses of cosmetic products. In *CMBEBIH*; Springer Press: Singapore, 2017; pp. 425–428.
10. Estanqueiro, M.; Amaral, M.H.; Sousa Lobo, J.M. Comparison between sensory and instrumental characterization of topical formulations: Impact of thickening agents. *Int. J. Cosmet. Sci.* **2016**, *38*, 389–398. [CrossRef] [PubMed]
11. Moravkova, T.; Filip, P. Relation between sensory analysis and rheology of body lotions. *Int. J. Cosmet. Sci.* **2016**, *38*, 558–566. [CrossRef] [PubMed]
12. Isaac, V.; Chiari, B.G.; Magnani, C.; Corrêa, M.A. Análise sensorial como ferramenta útil no desenvolvimento de cosméticos. *Rev. Ciênc. Farm. Básica Apl.* **2012**, *33*, 479–488.
13. Bourguet, C.; Pêcher, C.; Bardel, M.H.; Navarro, S.; Mougin, D. The use of an ethological approach to evaluate consumers' appreciation of luxury facial skincare and discriminate between products: A preliminary study. *Food Qual. Prefer.* **2016**, *50*, 7–14. [CrossRef]
14. Soto, M.L.; Conde, E.; González-López, N.; Conde, M.J.; Moure, A.; Sineiro, J.; Falqué, E.; Domínguez, H.; Núñez, M.J.; Parajó, J.C. Recovery and concentration of antioxidants from winery wastes. *Molecules* **2012**, *17*, 3008–3024. [CrossRef] [PubMed]

15. Conde, E.; Moure, A.; Domínguez, H.; Parajó, J.C. Production of antioxidants by non-isothermal autohydrolysis of lignocellulosic wastes. *LWT-Food Sci. Technol.* **2011**, *44*, 436–442. [CrossRef]
16. Fonseca-Santos, B.; Corrêa, M.A.; Chorilli, M. Sustainability, natural and organic cosmetics: Consumer, products, efficacy, toxicological and regulatory considerations. *Braz. J. Pharm. Sci.* **2015**, *51*, 17–26. [CrossRef]
17. Friberg, S.E. Vapour pressure of some fragrance ingredients in emulsion and microemulsion formulations. *Int. J. Cosmet. Sci.* **1997**, *19*, 75–86. [CrossRef] [PubMed]
18. Lee, C.J.; Chen, L.G.; Chang, T.L.; Ke, W.M.; Lo, Y.F.; Wang, C.C. The correlation between skin-care effects and phytochemical contents in Lamiaceae plants. *Food Chem.* **2011**, *124*, 833–841. [CrossRef]
19. Daudt, R.M.; Back, P.I.; Medeiros Cardozo, N.L.; Ferreira Marczak, L.D.; Kulkamp-Guerreiro, I.C. Pinhão starch and coat extract as new natural cosmetic ingredients: Topical formulation stability and sensory analysis. *Carbohydr. Polym.* **2015**, *134*, 573–580. [CrossRef] [PubMed]
20. Schifferstein, H.N.J.; Howell, B.F. Using color-odor correspondences for fragrance packaging design. *Food Qual. Prefer.* **2015**, *46*, 17–25. [CrossRef]

MDPI

St. Alban-Anlage 66

4052 Basel

Switzerland

Tel. +41 61 683 77 34

Fax +41 61 302 89 18

www.mdpi.com

Cosmetics Editorial Office

E-mail: cosmetics@mdpi.com

www.mdpi.com/journal/cosmetics

MDPI
St. Alban-Anlage 66
4052 Basel
Switzerland

Tel: +41 61 683 77 34
Fax: +41 61 302 89 18

www.mdpi.com

ISBN 978-3-03897-161-0

www.ingramcontent.com/pod-product-compliance
Lightning Source LLC
Chambersburg PA
CBHW051908210326
41597CB00033B/6064